Artificial Intelligence Techniques in IoT Sensor Networks

Chapman & Hall/CRC Distributed Sensing and Intelligent Systems Series

Series Editors: Mohamed Elhoseny and Xiaohui Yuan

Applications of Cloud Computing
Approaches and Practices
Prerna Sharma, Moolchand Sharma, Mohamed Elhoseny

Artificial Intelligence Techniques in IoT Sensor Networks
Mohamed Elhoseny, K. Shankar, Mohamed Abdel-Basset

For more information about this series please visit: https://www.routledge.com/Chapman—HallCRC-Distributed-Sensing-and-Intelligent-Systems-Series/book-series/DSIS

Artificial Intelligence Techniques in IoT Sensor Networks

Edited by
Mohamed Elhoseny
K. Shankar
Mohamed Abdel-Basset

CRC Press is an imprint of the
Taylor & Francis Group, an **informa** business
A CHAPMAN & HALL BOOK

First edition published 2021

by CRC Press
6000 Broken Sound Parkway NW, Suite 300, Boca Raton, FL 33487-2742

and by CRC Press
2 Park Square, Milton Park, Abingdon, Oxon, OX14 4RN

CRC Press is an imprint of Taylor & Francis Group, LLC

Library of Congress Cataloging-in-Publication Data

Names: Elhoseny, Mohamed, editor. | Shankar, K., editor. | Abdel-Basset, Mohamed, 1985- editor.
Title: Artificial intelligence techniques in IoT-sensor networks / edited by Mohamed Elhoseny, K. Shankar, Mohamed Abdel-Basset.
Description: First edition. | Boca Raton : CRC Press, 2021. | Series: Chapman & Hall/CRC distributed sensing and intelligent systems series | Includes bibliographical references and index.
Identifiers: LCCN 2020037221 | ISBN 9780367439255 (hardback) | ISBN 9781003007265 (ebook)
Subjects: LCSH: Sensor networks. | Internet of things.
Classification: LCC TK7872.D48 A78 2021 | DDC 006.3--dc23
LC record available at https://lccn.loc.gov/2020037221

ISBN: 978-0-367-43925-5 (hbk)
ISBN: 978-0-367-68145-6 (pbk)
ISBN: 978-1-003-00726-5 (ebk)

Typeset in Minion Pro
by KnowledgeWorks Global Ltd

Contents

Preface

IN RECENT TIMES, ARTIFICIAL intelligence (AI)-, machine learning (ML)-, and Internet of things (IoT)-based systems have become popular and find applicability in several domains. These solutions facilitate the creation of new products and services in many different fields. IoT architecture typically consists of a collection of sensors that gathers distinct kinds of data and transmits them to a "base station," which uploads the data to the cloud. From there, it can be made available to consumers and/or businesses as actionable intelligence. While there are IoT architectures that influence ubiquitous computing devices like smartphones as either "base station" or "sensor," the IoT-based sensor networks (SNs) allow the businesses to deploy IoT platforms and monitor the condition of their devices or other equipment. SNs are undergoing great expansion and development and the combination of both AI and SNs are now realities that are going to change our lives. The integration of these two technologies benefits other areas such as Industry 4.0, IoT, Demotic Systems, etc. Furthermore, SNs are widely used to collect environmental parameters in homes, buildings, vehicles, etc., where they are used as a source of information that aids the decision-making process and, in particular, it allows systems to learn and to monitor activity. New AI and ML real-time or execution time algorithms are needed, as well as different strategies to embed these algorithms in sensors.

This book, *Artificial Intelligence Techniques in IoT Sensor Networks*, examines the application of AI algorithms in the diverse aspects of SNs. This book includes original contributions on new methods and approaches to develop recent AI-enabled applications and intelligent decision-making models for SNs. It offers solutions to various real-time problems in SNs. This book contains 13 chapters. Topics range from healthcare diagnosis, wireless SNs, transportation, Social IoT (SIoT), face sketch synthesis (FSS), and so on, and are solely concentrated on AI applications.

Chapter 1 introduces AI-based IoT implementation for the teleradiology network for medical image segmentation using the adaptive regularized Gaussian kernel fuzzy c-means (FCM) technique. The chapter derives an improved FCM based on the adaptive regularized Gaussian kernel with the non-linear tensor diffusion filter for abdomen CT images.

Chapter 2 introduces AI- and IoT-based logistic transportation planning models based on the hybridization of fuzzy logic with the modified particle swarm optimization (HFMPSO) algorithm.

Chapter 3 develops a new big data analytics method in SIoT using butterfly optimization-based feature selection with the gradient boosting tree (GBT) technique. It also aims to employ the big data-enabled Hadoop framework for big data processing.

Chapter 4 proposes a new fuzzy logic-based clustering with data aggregation protocol for the wireless sensor networks (WSNs)-assisted IoT system. This method can be used to aggregate data using the error-bounded lossy compression (EBLC) technique.

Chapter 5 presents a smart home recommendation system from natural language processing services with a clustering technique. The proposed model aims to build a completely utilitarian voice-based home computerization framework using IoT, embedding (word displaying), and natural language processing services (NLPS).

Chapter 6 develops a metaheuristic-based kernel extreme learning machine (KELM) model for disease diagnosis in industrial IoT SNs. The presented model includes the spider monkey optimization (SMO) algorithm to fine-tune the parameters of KELM for maximized classification performance.

Chapter 7 introduces an effective new fuzzy support vector machine with the synthetic marginal over-sampling technique (SMOTE) model for the class imbalance problem in IoT and cloud-based disease diagnosis.

Chapter 8 presents an energy-efficient unequal clustering algorithm using hybridization of the social spider (SS) with krill herd (KH) in IoT-assisted WSN. The SS algorithm initially selects the tentative cluster heads and then the KH algorithm is applied to choose the final cluster heads.

Chapter 9 aims to develop an IoT- and 5G-enabled faster region proposal network with the fast regional convolutional neural network (RCNN) with the generative adversarial network (GAN) model for FSS.

Chapter 10 introduces a new AI-based textual cyberbullying detection for twitter data analysis in cloud-based IoT. The presented model uses oppositional grasshopper optimization with convolutional neural network (OGHOCNN) for the detection process.

Chapter 11 develops an energy-efficient quasi-oppositional krill herd algorithm-based clustering (QOKHC) protocol for IoT SNs. The proposed QOKHC algorithm incorporates the concept of quasi-oppositional-based learning (QOBL) in the KH algorithm to increase the convergence rate.

Chapter 12 proposes an effective SIoT model for malicious node detection in WSN. The proposed method makes use of exponential kernel (EK) procedures for distinguishing the malicious nodes in SIoT.

Chapter 13 devises an IoT-based automated skin lesion detection and classification using gray wolf optimization with deep neural network. The proposed method performs image segmentation, and the gray-level co-occurrence matrix (GLCM) technique is applied as a feature extractor.

About the Editors

Dr. Mohamed Elhoseny is assistant professor in the Department of Computer Science, College of Computer & Information Technology, American University in the Emirates (AUE). Dr. Elhoseny is an ACM Distinguished Speaker and IEEE Senior Member. He received his PhD in computers and information from Mansoura University/University of North Texas through a joint scientific program. Dr. Elhoseny is the founder and editor-in-chief of *IJSSTA*, published by IGI Global. Also, he is associate editor at *IEEE Journal of Biomedical and Health Informatics*, *IEEE Access*, *Scientific Reports*, *IEEE Future Directions*, *Remote Sensing*, and *International Journal of E-services and Mobile Applications*. Moreover, he served as the co-chair, publication chair, program chair, and track chair for several international conferences published by recognized publishers such as IEEE and Springer. Dr. Elhoseny is editor-in-chief of the Springer book series Studies in Distributed Intelligence, editor-in-chief of the forthcoming Taylor & Francis/CRC Press book series, Sensors Communication for Urban Intelligence and editor-in-chief of the Chapman & Hall/CRC Distributed Sensing and Intelligent Systems Series.

K. Shankar, member IEEE, is a postdoctoral fellow in the Department of Computer Applications, Alagappa University, Karaikudi, India. He has authored/co-authored over 54 ISI journal articles (with total impact factor 150+) and more than 100 Scopus indexed articles. He has guest-edited several special issues of many journals published by SAGE, TechScience, Inderscience, and MDPI. He has served as guest editor and associate editor for SCI, Scopus indexed journals like Elsevier, Springer, IGI, Wiley, and MDPI. He has served as chair (program, publications, technical committee, and track) on several international conferences. He has delivered several invited and keynote talks, and reviewed the technology leading articles for journals like *Scientific Reports—Nature*, *IEEE Transactions on Neural Networks and Learning Systems*, *IEEE Journal of Biomedical and Health Informatics*, *IEEE Transactions on Reliability*, *IEEE Access*, and *IEEE Internet of Things*. He has authored/edited conference proceedings and book chapters, and his two books have been published by Springer. He has been a part of various seminars, paper presentations, and research paper reviews. He has also served as convener and session chair of several conferences. He has displayed vast success in continuously acquiring new knowledge and applying innovative pedagogies and has always aimed to be an effective educator and have a global outlook. His current research interests include healthcare applications, secret image sharing scheme, digital image security, cryptography, IoT, and optimization algorithms.

Mohamed Abdel-Basset earned BSc, MSc, and PhD degrees in information systems and technology from the Faculty of Computers and Informatics, Zagazig University, Egypt. His current research interests are optimization, operations research, data mining, computational intelligence, applied statistics, decision support systems, robust optimization, engineering optimization, multiobjective optimization, swarm intelligence, evolutionary algorithms, and artificial neural networks. He is working on the application of multiobjective and robust metaheuristic optimization techniques. He is also editor/reviewer of different international journals and conferences. He has published more than 150 articles in international journals and conference proceedings. He has held program chair in many conferences in the fields of decision-making analysis, big data, optimization, complexity, and IoT, as well as editorial collaboration in journals of high impact.

CHAPTER 1

Adaptive Regularized Gaussian Kernel FCM for the Segmentation of Medical Images

An Artificial Intelligence-Based IoT Implementation for Teleradiology Network

S. N. Kumar,[1] A. Lenin Fred,[2] L. R. Jonisha Miriam[2], Ajay Kumar H.,[2] Parasuraman Padmanabhan,[3] and Balazs Gulyas[3]

[1]*Amal Jyothi College of Engineering, Kanjirappally, Kerala, India*

[2]*Mar Ephraem College of Engineering and Technology, Elavuvilai, Tamil Nadu, India*

[3]*Nanyang Technological University, Singapore*

CONTENTS

1.1 INTRODUCTION

Segmentation method is exploited for the extraction of desired region of interest (ROI) and in medical image processing, its role is pivotal in the delineation of anatomical organs and anomalies like tumor and cyst. There is no universal algorithm for various modalities, and the selection of segmentation technique relies on the type of imaging modality and ROI. Grouping of segmentation techniques relies on the nature of evolution and, in general, it is classified into semiautomatic and fully automatic. The semiautomatic algorithm requires human intervention: the discrete positioning of points in the level set model [1], selection of foreground, and background seed region in graph cut [2].

The big data analytics and Internet of Things (IoT) trends influence healthcare in the radiology sectors for the classification in efficient diagnosis [3]. A real-time mobile camera terminal captures the skin images that interact with the remote datacenter with a deep learning model, which improvises the learning model and predicts the skin disease classification and psychological depression [4]. For the analysis of medical imaging, the Raspberry Pi with sensors is used to collect data from the clinical environment by automatically logging telephone calls [5]. An ARM9-based processor was used for the intelligent processing of medical images, and the processed image was securely transferred with Li-Fi-enabled IoT [6]. The authors have reviewed the IoT applications in the E-healthcare by different types of sensors: medical big data management [7]. The author proposed a polynomial time algorithm for the efficient cloud data transfer to medical IoT devices with energy-efficient dynamic packet downloading of the medical images. This method adaptively changes power/energy at each access point with buffer stability [8]. The deep learning framework was proposed with Caffe or Tensor flow for training the machine learning model with medical images in a big data method. The trained model is used in mobile application for analysis and diagnosis of disease through cloud for Internet of Medical Imaging Thinks (IoMIT) [9]. The author introduced an edge computing device between the IoT device and cloud network for the effective analysis of CT brain images, especially for strokes patient. This model uses Adaptive Boosted Decision Trees (ABDT) and intelligent classifiers for the classification of cerebrospinal fluid (CSF), white matter (WM), gray matter (GM), ischemic stroke, hemorrhagic stroke, calcification, and bone for the better prediction of stroke recovery [10].

Fuzzy C-Means (FCM) is a soft clustering algorithm that groups the pixels with respect to fuzzy membership function. The random initialization of centroids, the prior setting of cluster number, and stuck at local minima are the discrepancies of classical FCM algorithm [11]. In general, the CT and MR images are corrupted by Gaussian and Rician noise, respectively. Moreover, appropriate filtering is required in the real-time scenario before any other image processing operations like segmentation, classification, and compression to improve accuracy [12]. Fuzzy clustering finds its application in medical and remote sensing areas [13, 14]. The traditional FCM initializes the cluster centroids arbitrarily, and it stucks in local optima regularly.

Initially, Dunn implemented FCM and it is modified by Hathaway and Bezdek [15, 16], whose objective function is given in equation (1.1).

$$J_{FCM} = \sum_{\alpha=1}^{N}\sum_{\beta=1}^{c} U_{\alpha\beta}^{m} d_{\alpha\beta}^{2} \tag{1.1}$$

where m is the fuzzy weighted exponential factor ($m > 1$), $d_{\alpha\beta}$ is the Euclidean distance, and $U_{\alpha\beta}$ indicates the membership function.

The Euclidean distance is represented in equation (1.2).

$$d_{\alpha\beta} = \left\| x_{\alpha} - V_{\beta} \right\|^{2} \tag{1.2}$$

The membership function satisfies the following constraint and is represented in equation (1.3):

$$\sum_{\beta=1}^{C} U_{\alpha\beta} = 1,\ U_{\alpha\beta} \epsilon [0,1] \tag{1.3}$$

Conventional FCM is subtle to noise due to the lack of local pixel information in objective function. The spatial FCM takes into account the neighborhood pixel information [17], whose objective function is expressed in equation (1.4):

$$J_{SFCM} = \sum_{\alpha=1}^{N}\sum_{\beta=1}^{C} U_{\alpha\beta}^{m} d_{\alpha\beta}^{2} + \frac{\gamma}{N_R} \sum_{\alpha=1}^{N}\sum_{\beta=1}^{C} U_{\alpha\beta}^{m} \sum_{r \epsilon N_{\alpha}} d_{r\beta}^{2} \tag{1.4}$$

γ controls the spatial information of the neighboring pixels with $0 > \gamma > 1$. N_{α} is the group of pixels and N_R is the cardinality of N_{α}.

The computational complexity of spatial fuzzy c-means (SFCM) is high and hence improved SFCM was proposed, which replaces $\frac{1}{N_R}\Sigma_{r \epsilon N_{\alpha}} d_{\beta}^{2}$ with $d^{2}\left(mean\left(x_{\alpha}, V_{\beta}\right)\right)$ [18] and is represented in equation (1.5):

$$J_{ESFCM1} = \sum_{\alpha=1}^{N}\sum_{\beta=1}^{C} U_{\alpha\beta}^{m} d_{\alpha\beta}^{2} + \gamma \sum_{\alpha=1}^{N}\sum_{\beta=1}^{C} U_{\alpha\beta}^{m} d^{2}\left(mean\left(x_{\alpha}, V_{\beta}\right)\right) \tag{1.5}$$

More time domain information is incorporated in objective function by replacing mean $\left(x_{\alpha}\right)$ by vector median of neighbours around x_{α} [19] and is expressed in equation (1.6):

$$J_{ESFCM2} = \sum_{\alpha=1}^{N}\sum_{\beta=1}^{C} U_{\alpha\beta}^{m} d_{\alpha\beta}^{2} + \gamma \sum_{\alpha=1}^{N}\sum_{\beta=1}^{C} U_{\alpha\beta}^{m} d^{2}\left(V_{m}\left(x_{\alpha}\right), V_{\beta}\right) \tag{1.6}$$

where $V_{m}\left(x_{\alpha}\right)$ is the vector median of image elements in the predefined kernel (usually 3×3).

The SFCM and enhanced spatial fuzzy c-means (ESFCM) require the estimation of parameters, and this problem was solved by fuzzy local information c-means (FLICM) algorithm [20]. The objective function of FLICM is represented in equation (1.7):

$$J_{FLICM} = \sum_{\alpha=1}^{N}\sum_{\beta=1}^{C} U_{\alpha\beta}^{m} d_{\alpha\beta}^{2} + G_{\alpha\beta} \tag{1.7}$$

$G_{\alpha\beta}$ is the fuzzy factor that incorporates both time domain and grayscale information of neighboring image elements. In every iteration, determination of fuzzy factor is needed, and hence FLICM algorithm is slow and also there is a loss of fine details of image due to the smoothing factor. The kernel weighted fuzzy local information c-means (KWFLICM) technique incorporates a weighted fuzzy factor in the objective function [21] and is represented in equation (1.8):

$$J_{KWFLICM} = \sum_{\alpha=1}^{N}\sum_{\beta=1}^{C} U_{\alpha\beta}^{m} d_{\alpha\beta}^{2} + G_{\alpha\beta} \tag{1.8}$$

where

$$G_{\alpha\beta} = \sum_{k \epsilon N_{\alpha},\, \alpha \neq k} w_{\alpha k}\left(1-U_{\alpha\beta}\right)^{m}\left(1-k\left(x_{\alpha}, V_{\beta}\right)\right)$$

The computation cost increases due to the trade-off weighted fuzzy factor and there is a loss of fine details of the image. Though FLICM and KWFLICM improve the segmentation result by the inclusion of time domain and grayscale information, the objective function is not minimized to an optimum value.

Hwang modified the FCM by replacing the conventional fuzzy membership function [22]. The expression $U_{\alpha\beta}$ was replaced by $a_{\alpha\beta}$ and is expressed in equation (1.9) as follows:

$$a_{\alpha\beta} = U_{\alpha\beta} - \frac{\left(1-U_{\alpha\beta}\right)}{2} \tag{1.9}$$

The type 2 FCM does not produce satisfactory output in the case of complex ROI patterns in the image. The objective function of type 2 FCM is represented in equation (1.10):

$$J_{TYPE2FCM} = \sum_{\alpha=1}^{N}\sum_{\beta=1}^{C} a_{\alpha\beta}^{m} d_{\alpha\beta}^{2} \tag{1.10}$$

It will terminate at a point, at which the previous and updated membership satisfy the following criteria represented in equation (1.11):

$$\max\left|a_{\alpha p}^{new} - a_{\alpha p}^{old}\right| < \rho \tag{1.11}$$

The novel variant of the FCM algorithm is institutionistic fuzzy c-means (IFCM) based on institutionistic fuzzy set [23]. Section 1.2 highlights the proposed methodology, Section 1.3 describes the results and discussion, and Section 1.4 constitutes the conclusion.

1.2 PROPOSED METHODOLOGY

For improving the efficiency of clustering technique, several approaches are proposed. The proposed clustering algorithm uses a regularization parameter for preserving the image details, partial differential equation-based filter to minimize the effect of Gaussian noise in CT images and Rician noise in MR images, and the Euclidean distance term in conventional clustering algorithm was replaced by a term comprising Gaussian radial basis function (GRBF) to achieve improved accuracy. The local context information was exploited by the introduction of the regularization term.

1.2.1 Fuzzy C-Means Clustering

FCM is a clustering technique that classifies a group of data into two or more cluster groups. It depends on the reduction of objective function [24]. The proposed algorithm involves the selection of the cluster center via optimization and segmentation by grouping data by cluster centers. The initial cluster points were optimally replaced by the optimization algorithm.

The squared distance from the pattern to the cluster center is illustrated as objective function, which is shown in equation (1.12):

$$J=\sum_{\alpha=1}^{P}\sum_{\beta=1}^{C}U_{\alpha\beta}^{f}d\left(y_{\alpha},V_{\beta}\right)=\sum_{\alpha=1}^{P}\sum_{\beta=1}^{C}U_{\alpha\beta}^{f}\left\|y_{\alpha},V_{\beta}\right\|^{2} \tag{1.12}$$

where C symbolizes the number of clusters, P is the number image elements, $f>1$ indicates the fuzzifier intensity, and $U_{\alpha\beta}$ is the membership function.

The membership function is represented in equation (1.13):

$$U_{\alpha\beta}=\frac{1}{\sum_{c=1}^{C}\left(\frac{D_{\alpha\beta}}{D_{\alpha c}}\right)^{\frac{2}{f-1}}}=\frac{1}{\sum_{c=1}^{C}\left(\frac{\left\|y_{\alpha}-V_{\beta}\right\|^{2}}{\left\|y_{\beta}-V_{c}\right\|^{2}}\right)^{\frac{1}{f-1}}} \tag{1.13}$$

The cluster center is updated in membership function as in equation (1.14):

$$V_{\beta}=\frac{\sum_{\alpha=1}^{N}U_{\alpha\beta}^{f}y_{\alpha}}{\sum_{\alpha=1}^{N}U_{\alpha\beta}^{f}} \tag{1.14}$$

1.2.2 Formulation of Nonlinear Tensor Diffusion Filtered Image

Generally CT and MR images are perverted by normal and Rician noise. The nonlinear tensor diffusion filter was found to be proficient in the filtering of normal and Rician noise [25]. The filtered image was used in the formulation of regularization term.

1.2.3 Improved Adaptive Regularized Kernel FCM

The regularization term is determined from the local variation coefficient (LCV), its value increases when there is high heterogeneity in the neighboring pixels:

$$LVC_\alpha = \frac{\sum_{P \in W_\alpha} \left(y_P - \overline{y}_\alpha\right)^2}{W_R * \left(\overline{y}_\alpha\right)^2} \tag{1.15}$$

where W_α is the grayscale of the pixel in the local window around the image element α, W_R is the cardinality of W_α, and $\overline{y}_\alpha$ is the mean grayscale.

The weights are determined from the exponential function framed from LVC_α.

$$W_\alpha = \frac{\zeta_\alpha}{\sum_{P \in W_\alpha} \zeta_P} \tag{1.16}$$

where

$$\zeta_\alpha = exp \sum_{P \in W_\alpha,\, \alpha \neq P} LVC_P$$

The regularization function is written as follows

$$\psi_\alpha = \begin{cases} 2 + W_\alpha, \ \overline{y}_i < y_\alpha \\ 2 - W_\alpha, \ \overline{y}_i > y_\alpha \\ 0, \ \overline{y}_i = y_\alpha \end{cases} \tag{1.17}$$

The pixels with a high value of LVC are assigned with a higher value of ψ_α. From equation (1.17), ψ_α value is 0, when the center pixel gray value of the local window is equal to the local window mean grayscale value. The parameter ψ_α is calculated in advance prior to the clustering process; hence, the computation time is significantly reduced. The computational complexity of clustering algorithms like FLICM, KWFLICM, T2FCM, and IFCM are high when compared with the proposed methodology. The regularization parameter explores the heterogeneity of pixel gray values; improved clustering result is produced since ψ_α is incorporated in the objective function.

The Euclidean distance term $\left\|y_\alpha - V_\beta\right\|^2$ can be replaced with $\left\|\psi\left(y_\alpha\right) - \psi\left(V_\beta\right)\right\|^2$, which is defined as follows:

$$\left\|\psi\left(y_\alpha\right) - \psi\left(V_\beta\right)\right\|^2 = F\left(y_\alpha, y_\alpha\right) + F\left(V_\beta, V_\beta\right) - 2F\left(y_\alpha, V_\beta\right) \tag{1.18}$$

where F denotes the kernel function.

In this work, GRBF kernel is used

$$F\left(y_{\alpha},V_{\beta}\right)=exp\left(\frac{-\left\|y_{\alpha}-V_{\beta}\right\|^{2}}{2\sigma^{2}}\right) \tag{1.19}$$

where σ is the kernel width.

In terms of GRBF kernel function, equation (1.18) is defined as follows:

$$\left\|\psi\left(y_{\alpha}\right)-\psi\left(V_{\beta}\right)\right\|^{2}=2\left(1-F\left(y_{\alpha},V_{\beta}\right)\right) \tag{1.20}$$

The value of σ plays a vital role and is determined as follows:

$$\sigma=\left[\frac{\sum_{\alpha=1}^{N}\left(k_{\alpha}-\bar{k}\right)^{2}}{N-1}\right]^{1/2} \tag{1.21}$$

where $k_{\alpha}=\left\|y_{\alpha}-\bar{y}\right\|$ is the distance from the gray level of image element α to the gray level average of all image elements and $\bar{y}$ is the average of all distances k_{α}.

The improved Adaptive Regularized Kernel-based Fuzzy Clustering Means (ARKFCM) objective function in terms of the GRBF kernel and regularization term is as follows:

$$J_{IARKFCM}=2\left[\sum_{\alpha=1}^{N}\sum_{\beta=1}^{c}U_{\alpha\beta}^{f}\left(1-F\left(y_{\alpha},V_{\beta}\right)\right)+\sum_{\alpha=1}^{N}\sum_{\beta=1}^{c}\psi_{\alpha}U_{\alpha\beta}^{f}\left(1-F\left(\bar{y}_{\alpha},V_{\beta}\right)\right)\right] \tag{1.22}$$

$$U_{\alpha\beta}=\frac{\left(\left(1-F\left(y_{\alpha},V_{\beta}\right)\right)+\psi_{\alpha}\left(1-F\left(\bar{y}_{\alpha},V_{\beta}\right)\right)\right)^{\frac{-1}{(f-1)}}}{\sum_{k=1}^{c}\left(\left(1-F\left(y_{\alpha},V_{\beta}\right)\right)+\psi_{\alpha}\left(1-F\left(\bar{y}_{\alpha},V_{k}\right)\right)\right)^{\frac{-1}{(f-1)}}} \tag{1.23}$$

$$V_{\beta}=\frac{\sum_{\alpha=1}^{N}U_{\alpha\beta}^{f}\left(F\left(y_{\alpha},V_{\beta}\right)y_{\alpha}+\psi_{\alpha}F\left(\bar{y}_{\alpha},V_{\beta}\right)\bar{y}_{\alpha}\right)}{\sum_{\alpha=1}^{N}U_{\alpha\beta}^{f}\left(F\left(y_{\alpha},V_{\beta}\right)+\psi_{\alpha}F\left(\bar{y}_{\alpha},V_{\beta}\right)\right)} \tag{1.24}$$

The improved ARKFCM generates efficient results than the classical clustering algorithms. The incorporation of nonlinear tensor diffusion filtered images improves the noise immunity.

1.3 RESULTS AND DISCUSSION

Clustering techniques are implemented using MATLAB 2015a and validated on real-time abdomen datasets. Seven CT abdomen datasets are used, the output of typical slices is illustrated here. Apart from the proposed clustering methodology, three more prominent clustering algorithms are used for the comparative analysis: T2FCM, FCICM, and SFCM.

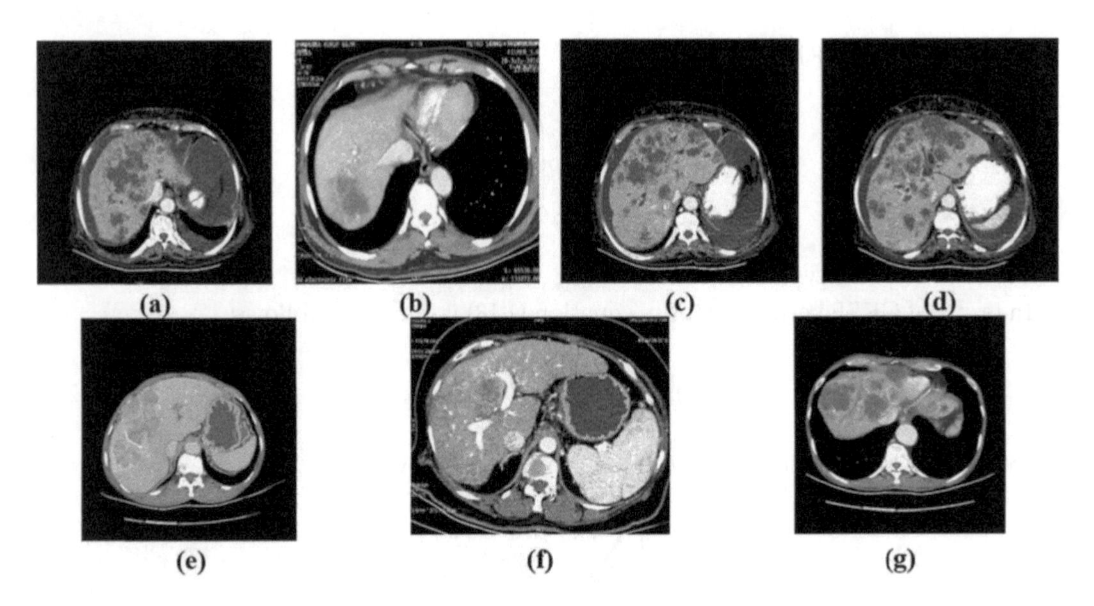

FIGURE 1.1 DICOM input images (ID1–ID7).

Initially, the performance metrics like Partition Coefficient (PC) and Partition Entropy (PE) were used to validate the efficiency of clustering algorithm, the proposed methodology was found to have lower PE and high PC, which indicates its proficiency. The proposed clustering technique was then validated for the choice of clustering number selection by the following metrics. The performance metrics validation reveals that the proposed methodology is efficient for C = 3 and the results are tabulated below. The input images are depicted in Figure 1.1. Figure 1.2 depicts the results of ID1 for different clustering algorithms (SFCM, FLICM, T2FCM, and improved ARKFCM). The PC and PE values indicate that improved

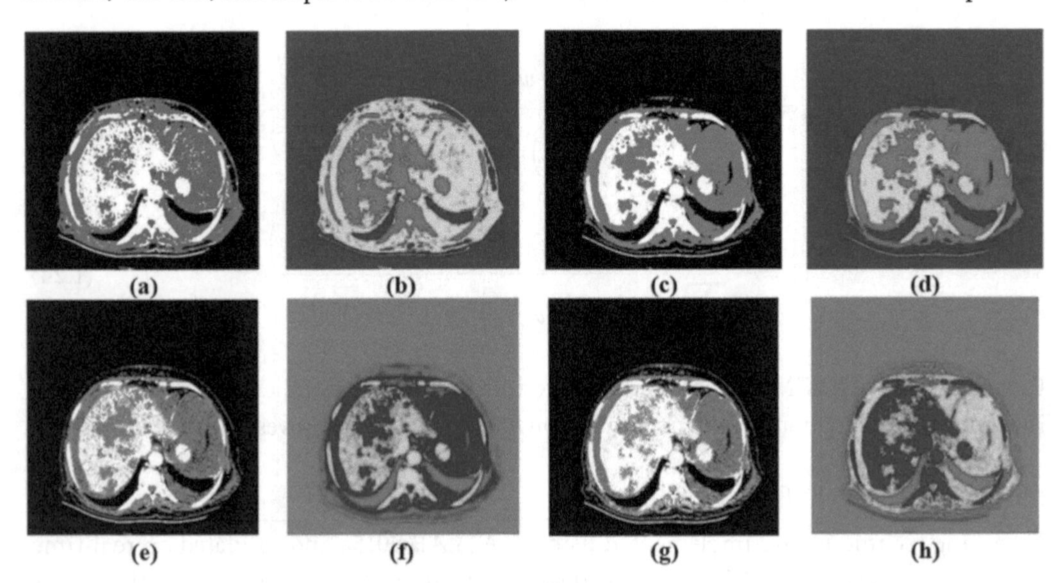

FIGURE 1.2 ID1 outputs corresponding to different clustering algorithms: SFCM, FLICM, T2FCM, and ARKFCM.

TABLE 1.1 Cluster Validity Index Set 1

Cluster Validity Index	Formula
Partition coefficient (PC) [26, 27]	$PC=\frac{1}{N}\sum_{p=1}^{P}\sum_{\alpha=1}^{N}\mu_{\alpha p}^{2}$
Partition entropy (PE) [26, 27]	$PE=\frac{1}{N}\sum_{p=1}^{P}\sum_{\alpha=1}^{N}\mu_{\alpha p}\log_2\left(\mu_{\alpha p}\right)$
Xie and Beni index (XBI) [28]	$XBI=\frac{\sum_{p=1}^{P}\sum_{\alpha=1}^{N}\mu_{\alpha p}^{K}\left\Vert u_{\alpha}-v_{p}\right\Vert^{2}}{N.\min_{\alpha\neq\beta}\left\Vert v_{\alpha}-v_{\beta}\right\Vert^{2}}$
The Fukuyama and Sugeno index (FSI) [28]	$FSI=\sum_{p=1}^{P}\sum_{\alpha=1}^{N}\mu_{\alpha p}^{f}\left\Vert u_{\alpha}-v_{p}\right\Vert^{2}-\sum_{p=1}^{P}\sum_{\alpha=1}^{N}\mu_{\alpha p}^{f}\left\Vert v_{p}-\hat{v}\right\Vert^{2}$

ARKFCM is superior, when compared with the other clustering algorithms. The improved ARKFCM clustering results for various cluster values for ID1 are depicted in Figure 1.3. The improved ARKFCM clustering outputs for ID2–ID6 is depicted in Figures 1.4–1.6. The outputs are taken for various cluster values ranging from C = 3 to C = 6.

Figures 1.7 and 1.8 depict the performance plot of PC and PE corresponding to different clustering algorithms, respectively.

Tables 1.1 and 1.2 represent the cluster validity index sets 1 and 2, respectively.

For selecting cluster values for segmentation, the metrics depicted in Table 1.2 are also used. For efficient clustering algorithm, the + and – signs symbolize that the metric value must be high and low, respectively. Tables 1.3 and 1.4 depict the cluster validity index sets 1 and 2, respectively, for the datasets (ID1–ID7).

TABLE 1.2 Cluster Validity Index Set 2

Cluster Validity Index	Formula
Calinski-Harabasz index (CHI) [29]	$CHI=\frac{B_{P}}{P-1}/\frac{W_{P}}{N-P}$
Silhouette coefficient index (SCI) [29]	$SCI=SC_{1}-SC_{2}$
Davies Bouldin index (DBI) [30]	$DBI=\frac{1}{P}\sum_{p=1}^{P}\max_{\beta\neq p}\frac{S_{\beta}+S_{p}}{\left\Vert V_{\beta}-V_{p}\right\Vert^{2}}$
Partition coefficient and exponential separation index (PCAESI) [29]	$PCAESI=\sum_{p=1}^{P}\sum_{\alpha=1}^{N}\frac{\mu_{\alpha p}^{2}}{\mu_{M}}-exp\left(\frac{-\min_{h\neq p}\left\Vert V_{p}-V_{h}\right\Vert^{2}}{\rho_{T}}\right)$
Pakhira-Bandyopadhyay-Maulik index (PBMFI) [30]	$PBMFI=\frac{\max_{\beta\neq lp}\left\Vert V_{\beta}-V_{p}\right\Vert\sum_{\alpha=1}^{N}\mu_{\alpha l}\left\Vert U_{\alpha}-V_{l}\right\Vert}{P\sum_{p=1}^{P}\sum_{\alpha=1}^{N}\mu_{\alpha p}^{f}\left\Vert U_{\alpha}-V_{p}\right\Vert}$
WL index (WLI) [29]	$WLI=\sum_{p=1}^{P}\sum_{\alpha=1}^{N}\frac{\mu_{\alpha p}^{2}\left\Vert U_{\alpha}-V_{p}\right\Vert^{2}}{\sum_{\alpha=1}^{N}\mu_{\alpha P}}$

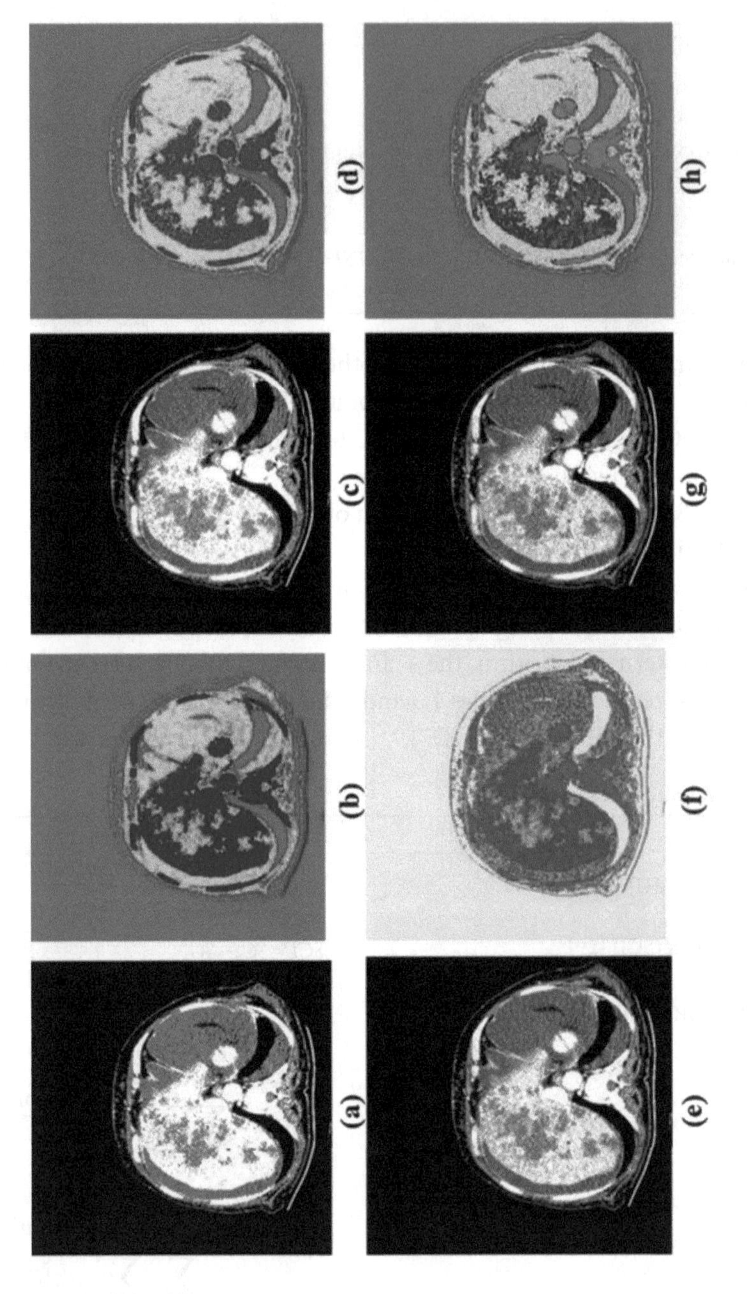

FIGURE 1.3 ID1 outputs of improved ARKFCM algorithm for various cluster values: C = 3, 4, 5, and 6.

TABLE 1.3 Cluster Validity Index Set 1 for the Datasets (ID1–ID7)

	Datasets with Chosen Cluster Values						
Cluster Validity Set 1	ID1 C = 3	ID2 C = 4	ID3 C = 3	ID4 C = 3	ID5 C = 4	ID6 C = 4	ID7 C = 3
CHI+	131070	161022	131070	131070	131070	644094	131070
DBI–	3.4123	2.5605	3.2604	3.5277	9.2182	4.0362	10.0691
XBI–	62139.4866	84440	65763	64984	127010	63762	121887
SCI+	−2.6077	−2.5555	−2.5931	−2.5798	−2.6314	−2.5416	−2.6310
PBM+	763	1806	952	1159	39967	2377	240

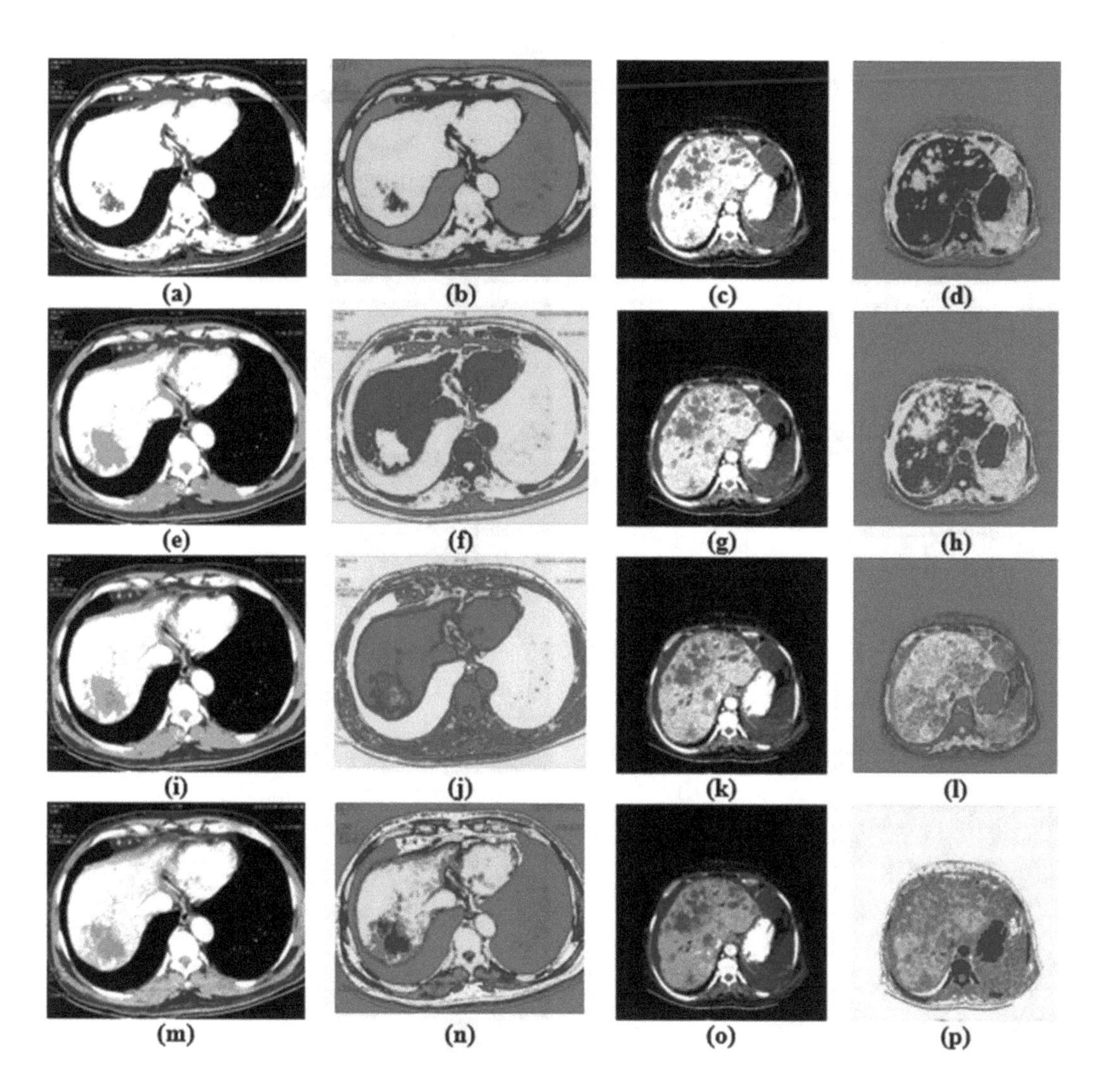

FIGURE 1.4 ID2 and ID3 outputs of improved ARKFCM algorithm for various cluster values: C = 3, 4, 5, and 6 in each row.

TABLE 1.4 Cluster Validity Index Set 2 for the Datasets (ID1–ID7)

	Datasets with Chosen Cluster Values						
Cluster Validity Set 2	**ID1** C = 3	**ID2** C = 4	**ID3** C = 3	**ID4** C = 3	**ID5** C = 4	**ID6** C = 4	**ID7** C = 3
FSI–	21987	8985	6901	6859	1555	2642	2957
PCA+	95540	5507	6238	8955	2292	1442	7770
WLI–	68073	9425	7344	7617	1170	8070	9933
Inter Cluster distance +	1.33 e+009	1.38 e+009	1.38 e+009	1.52 e+009	9.43 e+008	1.78 e+009	7.37 e+008
Intra Cluster distance –	1.012 e+009	2.4 e+009	1.168 e+009	1.27 e+009	3.85 e+009	2.5 e+009	1.01 e+009

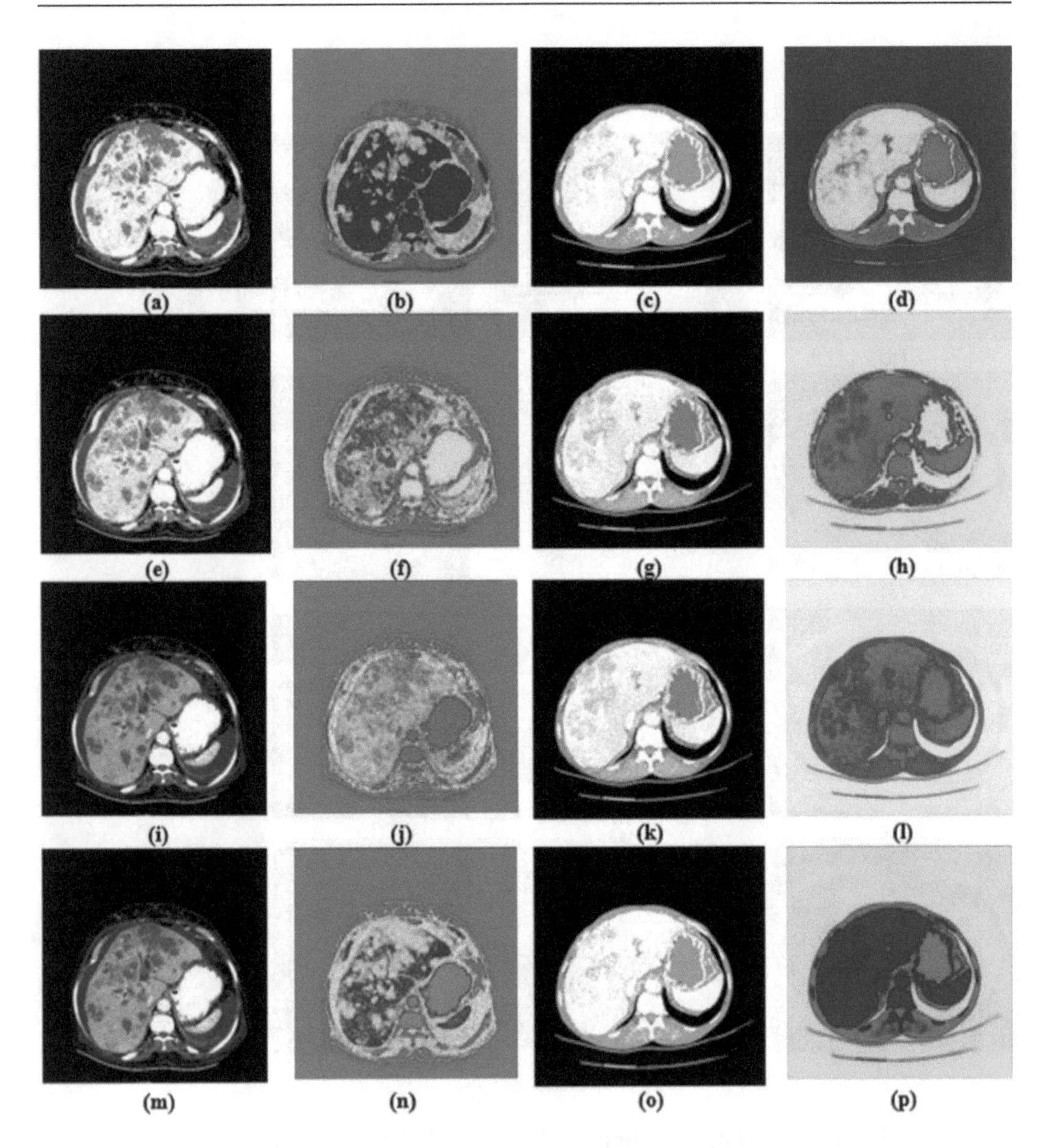

FIGURE 1.5 ID4 and ID5 outputs of improved ARKFCM algorithm for various cluster values: C = 3, 4, 5, and 6 in each row.

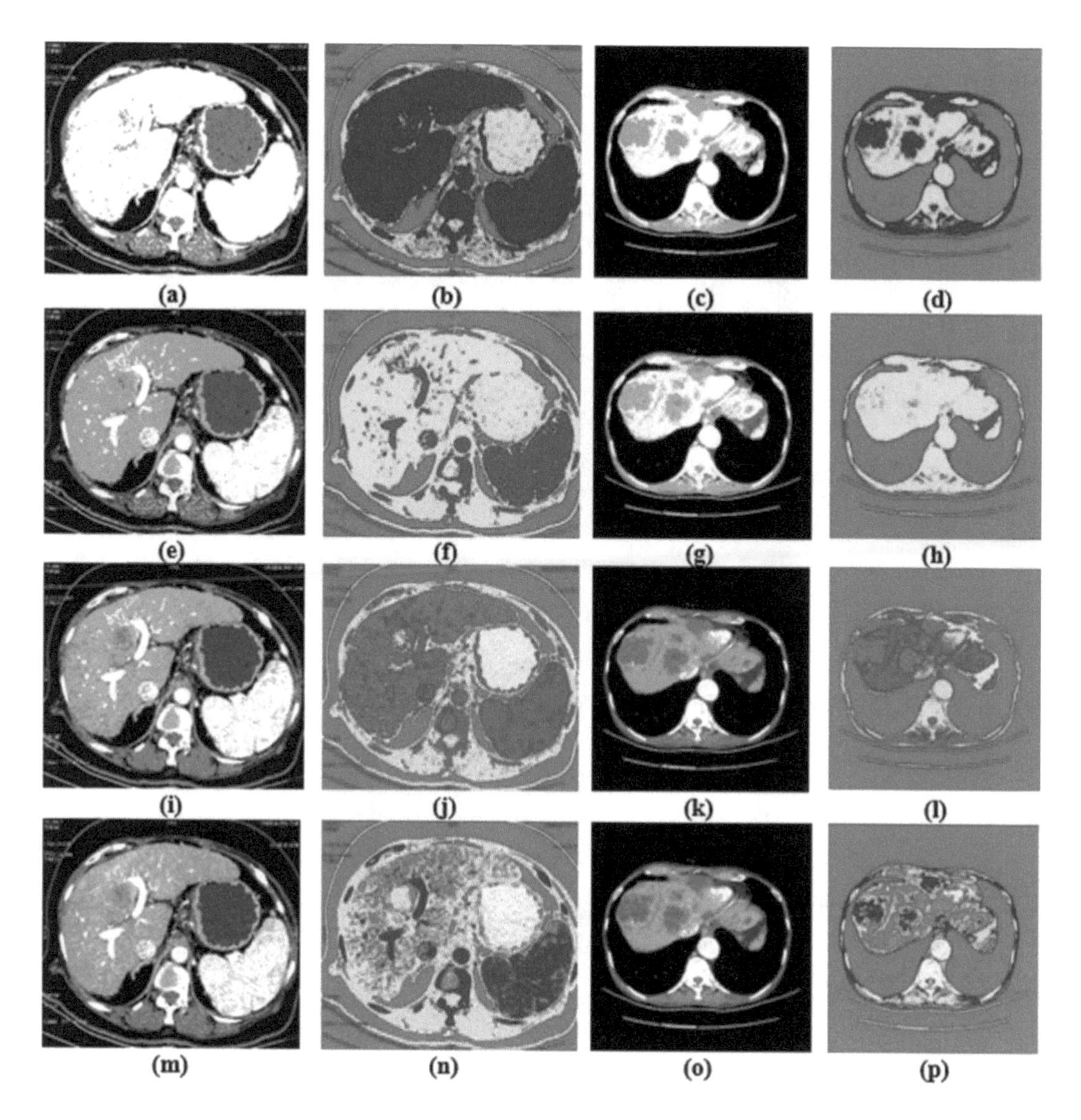

FIGURE 1.6 ID6 and ID7 outputs of improved ARKFCM algorithm for various cluster values.

For hardware implementation, the latest embedded processor Raspberry Pi 3 B+ model was used and the programming was done in Java. It is a 64-bit quad core processor that works at 1.4 GHz. It has built-in Wi-Fi and Ethernet. A user-friendly graphical user interface (GUI) was developed for loading DICOM image, performing clustering operation, saving segmentation result, and transferring data through cloud network. The hardware implementation is depicted in Figure 1.9.

Figure 1.10 depicts the GUI model for the segmentation system. DICOM and HFC format medical images are loaded to perform desired operations. Once the operations are performed, the results are saved and can be transferred to the desired nodes through the cloud network. Future works focus on the usage of optimization technique for the optimum selection of clusters and for hardware implementation, improved processor architecture with parallel processing can be employed for fast data transfer.

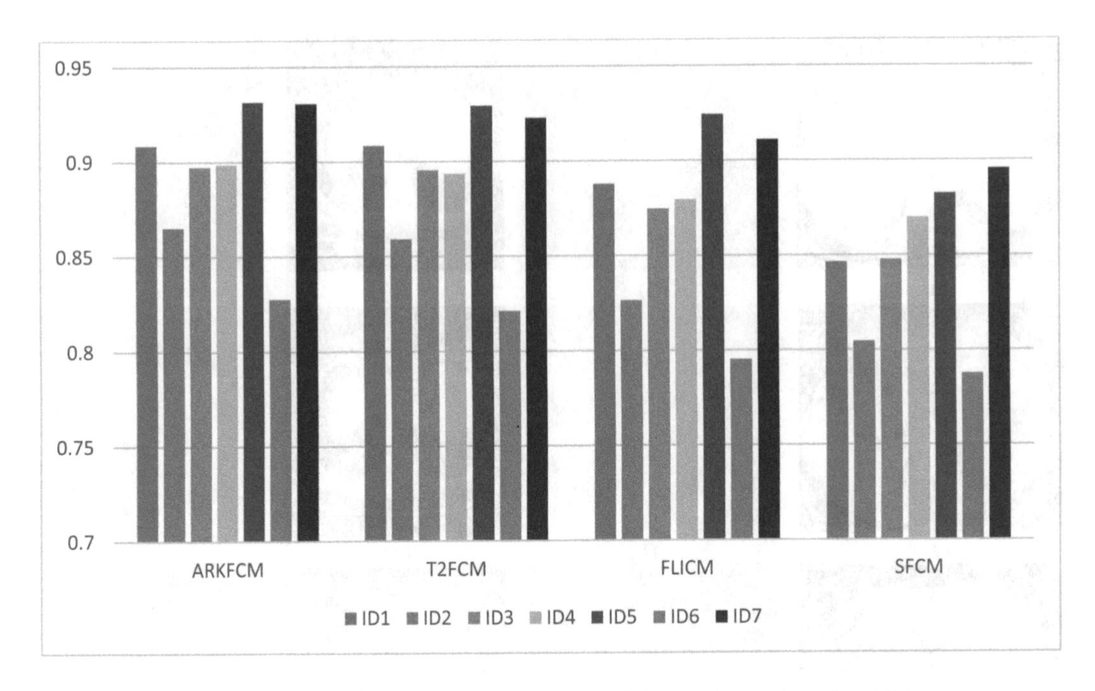

FIGURE 1.7 Performance plot of PC corresponding to different clustering algorithms.

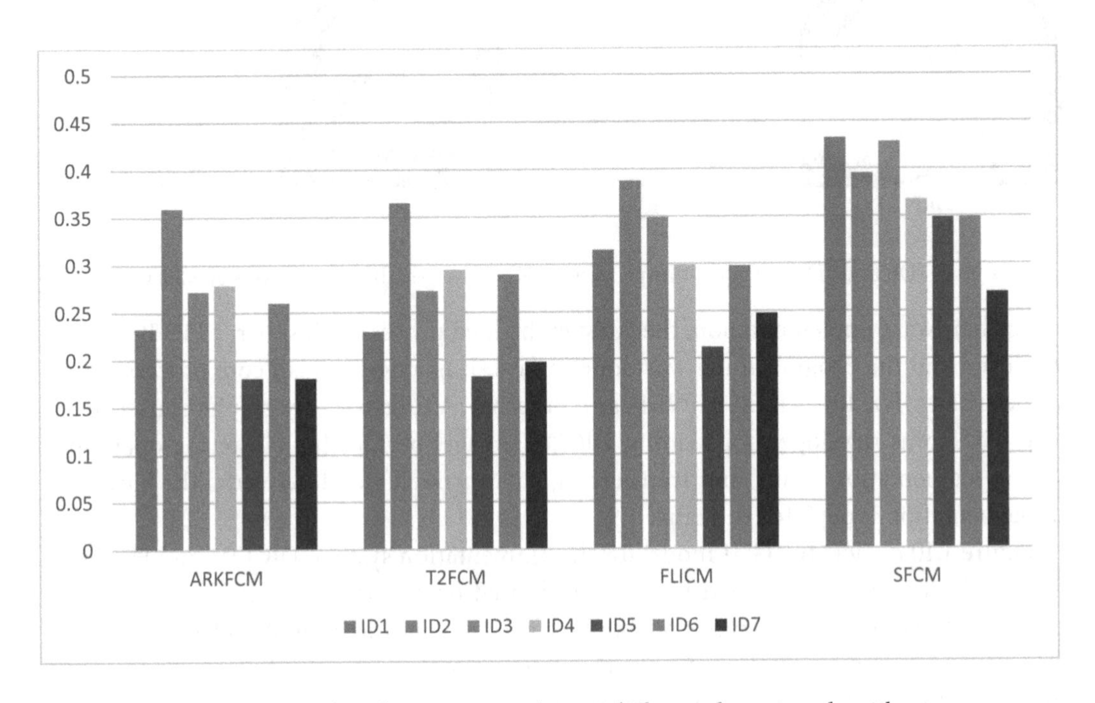

FIGURE 1.8 Performance plot of PE corresponding to different clustering algorithms.

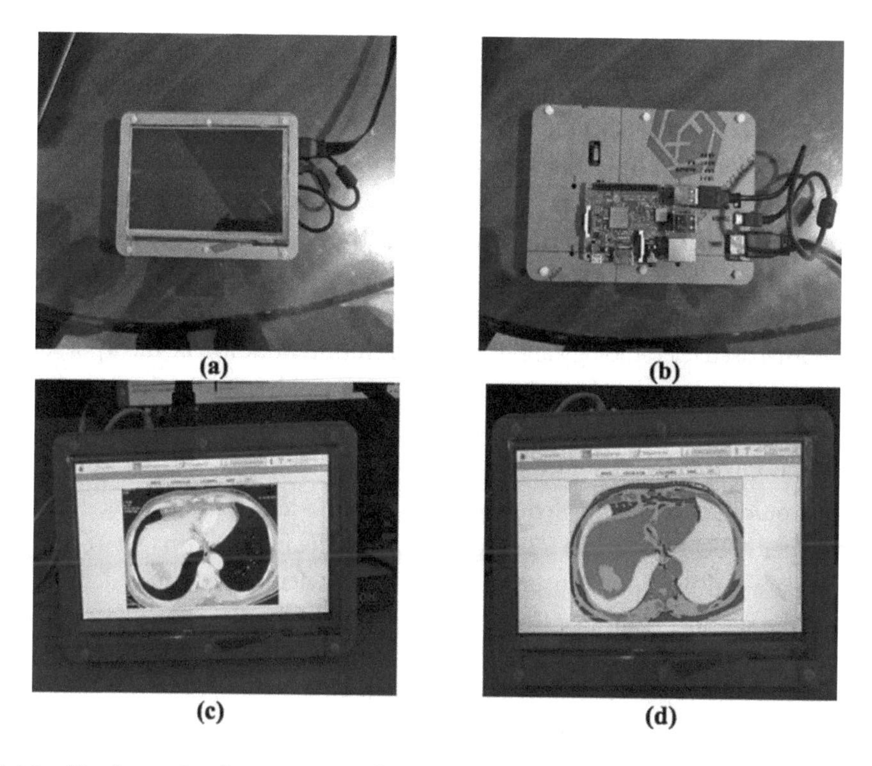

FIGURE 1.9 Hardware implementation of segmentation algorithm in Raspberry Pi embedded board.

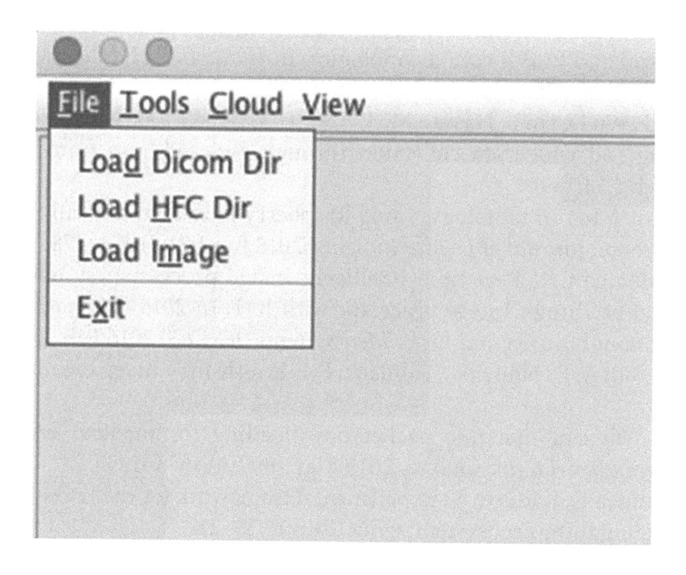

FIGURE 1.10 GUI developed for the segmentation system.

1.4 CONCLUSION

This chapter proposes a novel clustering approach for the segmentation of medical images. The improved regularized adaptive Gaussian kernel FCM with nonlinear tensor diffusion filter was used for the extraction of ROI. The incorporation of filter in the clustering approach makes it less sensitive to noise. The proposed improved ARKFCM generates efficient clustering result, when compared with the other clustering approaches. The performance validation was done by clustering metrics and the results were compared with the other clustering approaches; robust results were produced. The hardware implementation was carried out in the Raspberry Pi B^+ processor for IoT-based teleradiology application. The future work will be incorporation of bio-inspired optimization algorithm for automatic cluster centroids initialization.

ACKNOWLEDGMENTS

The authors would like to acknowledge the support provided by Nanyang Technological University under NTU Ref: RCA-17/334 for providing the medical images and supporting us in the preparation of the manuscript. Parasuraman Padmanaban and Balazs Gulyas acknowledge the support from Lee Kong Chian School of Medicine and Data Science and AI Research Centre (DSAIR) of NTU (Project Number ADH-11/2017-DSAIR) and the support from the Cognitive NeuroImaging Centre (CONIC) at NTU.

REFERENCES

1. Rusnell BJ, Pierson RA, Singh J. et al. Level set segmentation of bovine corpora lutea in *ex situ* ovarian ultrasound images. *Reproductive Biology and Endocrinology.* 2008;6:33. https://doi.org/10.1186/1477-7827-6-33
2. Wu W, Zhou Z, Wu S, Zhang Y. Automatic liver segmentation on volumetric CT images using super voxel-based graph cuts. *Computational and Mathematical Methods in Medicine.* 2016;2016: 9093721.
3. Banerjee A, Chakraborty C, Kumar A, Biswas D. Emerging trends in IoT and big data analytics for biomedical and health care technologies. In *Handbook of data science approaches for biomedical engineering*, 2020 Jan 1 (pp. 121–152). Academic Press, Cambridge, MA.
4. Chen M, Zhou P, Wu D, Hu L, Hassan MM, Alamri A. AI-skin: skin disease recognition based on self-learning and wide data collection through a closed-loop framework. *Information Fusion.* 2020 Feb 1;54:1–9.
5. Chen PH, Cross N. IoT in radiology: using Raspberry Pi to automatically log telephone calls in the reading room. *Journal of Digital Imaging.* 2018 Jun 1;31(3):371–378.
6. Dinesh M, Sudhaman K. Real time intelligent image processing system with high speed secured Internet of Things: Image processor with IOT. In *2016 International Conference on Information Communication and Embedded Systems (ICICES)* 2016 Feb 25 (pp. 1–5). IEEE.
7. Scarpato N, Pieroni A, Di Nunzio L, Fallucchi F. E-health-IoT universe: a review. *Management.* 2017;21(44):46.
8. Kim J. Energy-efficient dynamic packet downloading for medical IoT platforms. *IEEE Transactions on Industrial Informatics.* 2015 May 18;11(6):1653–1659.
9. Borovska P, Ivanova D, Kadurin V. Experimental framework for the investigations in internet of medical imaging things ecosystem. *QED.* 2018;17:20–21.
10. Vasconcelos FF, Sarmento RM, Rebouças Filho PP, de Albuquerque VH. Artificial intelligence techniques empowered edge-cloud architecture for brain CT image analysis. *Engineering Applications of Artificial Intelligence.* 2020 May 1;91:103585.

11. Chuang KS, Tzeng HL, Chen S, Wu J, Chen TJ. Fuzzy c-means clustering with spatial information for image segmentation. *Computerized Medical Imaging and Graphics.* 2006 Jan 1; 30(1):9–15.
12. Trinh DH, Luong M, Rocchisani JM, Pham CD, Dibos F. Medical image denoising using kernel ridge regression. In *2011 18th IEEE International Conference on Image Processing* 2011 Sep 11 (pp. 1597–1600). IEEE.
13. Kannan SR, Ramathilagam S, Sathya A, Pandiyarajan R. Effective fuzzy c-means based kernel function in segmenting medical images. *Computers in Biology and Medicine.* 2010;40(6):572–579.
14. Atkinson PM. Sub-pixel target mapping from soft-classified, remotely sensed imagery. *Photogrammetric Engineering & Remote Sensing.* 2005;71(7):839–846.
15. Dunn, J.C. A fuzzy relative of the isodata process and its use in detecting compact well-separated clusters. *Journal of Cybernetics.* 1973;3:32–57.
16. Hathaway, R.J., Bezdek, J.C. Recent convergence results for the fuzzy c-means clustering algorithms. *Journal of Classification.* 1988;5(2):237–247.
17. Van L, Koen FM, Dirk V, Paul S. Automated model-based tissue classification of MR images of the brain. *IEEE Transactions on Medical Imaging.* 1999;18:897–908.
18. Abdullah A, Hirayama A, Yatsushiro S, Matsumae M, Kuroda K. Cerebrospinal fluid image segmentation using spatial fuzzy clustering method with improved evolutionary expectation maximization. *IEEE 35th Annual International Conference on Medicine and Biology* 2013 Jul 3 (pp. 3359–3362). Osaka, Japan.
19. Cai W, Chen S, Zhang D. Fast and robust fuzzy c-means clustering algorithms incorporating local information for image segmentation. *Pattern Recognition.* 2007;40(3):825–38.
20. Krinidis S, Vassilios C. A robust fuzzy local information c-means clustering algorithm. *IEEE Transactions on Image Processing.* 2010;19:1328–1337.
21. Shalini R, Muralidharan V, Varatharaj M. MRI brain tumor segmentation using kernel weighted fuzzy clustering. *International Journal of Engineering Research and Technology.* 2014;3(4):121–125.
22. Aneja D, Tarun KR. Fuzzy clustering algorithms for effective medical image segmentation. *International Journal of Intelligent Systems Application.* 2013;11:55–61.
23. Han J, Yang Z, Sun X, Xu G. Chordal distance and non-Archimedean chordal distance between Atanassov's intuitionistic fuzzy set. *Journal of Intelligent & Fuzzy Systems.* 2017;33(6):3889–3994.
24. Xinbo G, Jie L, Weixin X. Parameter optimization in FCM clustering algorithms. In *WCC 2000-ICSP 2000. 2000 5th International Conference on Signal Processing Proceedings. 16th World Computer Congress* 2000 (Vol. 3, pp. 1457–1461). IEEE.
25. Kumar SN, Fred AL, Kumar HA, Varghese PS. Nonlinear tensor diffusion filter based marker-controlled watershed segmentation for CT/MR images. In *Proceedings of International Conference on Computational Intelligence and Data Engineering* 2018 (pp. 317–331). Springer, Singapore.
26. Wu KL, Yang MS. A cluster validity index for fuzzy clustering. *Pattern Recognition Letters.* 2005 Jul 1;26(9):1275–1291.
27. Pal NR, Bezdek JC. On cluster validity for the fuzzy c-means model. *IEEE Transactions on Fuzzy Systems.* 1995 Aug;3(3):370–379.
28. Kim DW, Lee KH, Lee D. Fuzzy cluster validation index based on inter-cluster proximity. *Pattern Recognition Letters.* 2003 Nov 1;24(15):2561–2574.
29. Wu CH, Ouyang CS, Chen LW, Lu LW. A new fuzzy clustering validity index with a median factor for centroid-based clustering. *IEEE Transactions on Fuzzy Systems.* 2014 May 7;23(3):701–718.
30. Pakhira MK, Bandyopadhyay S, Maulik U. Validity index for crisp and fuzzy clusters. *Pattern Recognition.* 2004 Mar 1;37(3):487–501.

CHAPTER 2

Artificial Intelligence-Based Fuzzy Logic with Modified Particle Swarm Optimization Algorithm for Internet of Things-Enabled Logistic Transportation Planning

Phong Thanh Nguyen

Department of Project Management, Faculty of Civil Engineering, Ho Chi Minh City Open University, Ho Chi Minh City, Vietnam

CONTENTS

2.1 INTRODUCTION

Logistics are modified similar to the changes in retail sales. Prior to using e-commerce, the establishment of logistics was classified into three stages according to the alterations in logistics providers [1]. Industrial Revolution has resulted in several manufacturing organizations, which have been modeled in a way as to manage the record as well as transportation of the products. The economic globalization that promotes social division has led to the practice of outsourcing of logistics by producers and sellers to the third-party logistics firms having expertise in the field. This reduces the costs and improves efficiency. With e-commerce, the tasks of logistics have undergone change, serving as links between production and sales by delivering consumables directly to the users. For example, the network-based delivery organizations as well as warehouse and logistics enterprises are significant links between producers and clients [2].

Big data analytics is a technique that is applied to define a massive amount of data collection with respect to acquisition, memory, management, and analysis; this technique exceeds the abilities of existing database software tools [3]. Big data has been characterized by high-volume data, rapid information flow, variety of data, and minimum rate of density [4]. For example, number space has been employed by the databases of social network site Facebook per day; the New York Stock Exchange produces 1 TB of novel trade data in a single day; it generates an individual jet engine that is capable of producing 10 TB of data within a limited time frame. The importance of big data model can be estimated by processing the data that is used to analyze the included values instead of processing massive data. In various cases, with adequate data, a system could be developed with an application technique that guides the machines to perform smart objectives under the applications of different machine learning (ML) techniques [5].

The digitalization of logistics would be enhanced vitally where several models of labor division, like crowd sourcing and crowd funding and sharing, are applied widely. The service economy and experience economy are improved further, as AI models are being evolved in a rapid manner; thus, the "Intelligent Revolution" would modify the logistics industry. Big data is more rapid in transforming business formats and lifestyle modifications, and helps in social and economic development trajectories, which is embedded with maximum positive impact on logistics industry. With respect to the application as well as industrial developments of methodologies, existing logistics are simplified by longer multiple link chains, automatic works, and locally optimized deployments that acquire immediate modifications to face the issues caused by the modifications in a market platform for developing logistics industry [6]. In last decades, a product has been launched and used by the users, which consumes major steps at the time of selling at least five products on average. Based on the research work, to reach the customer, processing and manufacturing should be of lower duration, whereas managing and transportation should be optimum. A longer chain as well as several links of classical logistics makes complex adjustments. The issues arising at the time of massive as well as longer processes tend to minimize efficiency and maximize expenses.

In the present-day business scenario, the deployment of models like big data, AI, and robotics tend to stimulate the basic modifications in logistics intelligence. Logistics systems and its applications lead to independent route development, exploring human visual system, and some other events, which combine the modern devices with diverse links. The actual decision-making process depends upon the experience of entirely converting AI model, where the systems attain self-thinking and autonomous decision-making action [7].

The main aim of logistics is to combine modern and professional systems that lead to developing logistics route planning, which enables reaching expert level through domain information as well as the inference engine. The traditional methods are referred to be representative techniques in evolutionary models like genetic algorithms (GAs) [8], ant colony optimization (ACO), artificial bee colony (ABC), swarm optimization (SO) [9], and so on.

According to the definition provided, the purpose of logistics route planning is to effectively explore for a logistics solution that may meet every logistics operation, such as pickup and delivery needs, and reduce the overall computational cost of logistics. In recognizing the model, the initial challenge is to define superiority of a logistics solution. It has been assumed that computational cost of logistics includes courier cost based on the working time and fuel cost of overall driving distance. Therefore, if the personnel and routing costs are less, then an optimized logistics solution is attained. On the other hand, as logistics route planning issue comes under the NP-hard issue, the scalability of a logistics platform increases the processing cost. Hence, the model of managing processing efficiency as well as solution superiority is assumed to be an alternative challenge. A logistics firm has longer duration to determine the maximum-quality logistics solution. Therefore, a real-time logistics planning technique must ensure a better practical solution; however, solution may further be enhanced on the advice of modern experts, if a logistics industry has maximum duration to obtain optimal solutions.

In this chapter, a new intelligent logistic transportation planning model is presented by the hybridization of fuzzy logic with modified particle swarm optimization (HFMPSO) algorithm. The proposed model initially used Internet of Things (IoT) based barcode reader to access the details of the package. The proposed transportation planning algorithm involves three main phases: partitioning of packages, route planning using HFMPSO algorithm, and package insertion. The HFMPSO model has been tested using a set of performance measures under diverse aspects. The experimental outcome clearly verified the superior performance of the HFMPSO model over the compared methods in a significant way.

2.2 RELATED WORKS

For solving the vehicle routing problem (VRP) in an effective fashion, a large number of heuristic approaches have been presented for logistics routes. Tabu search models as well as simulated annealing (SA) are employed by Osman [10]. The variable neighborhood search (VNS) and iterated local search (ILS) have been presented in [11]; the effective models are used to attain better solutions to enhance the primary route plan. Sze et al. [12] projected an adaptive VNS (AVNS) technique, which integrates large neighborhood search (LNS) in the form of diversification principle to use in a capable VRP. Miranda-Bront et al. [13] assumed a cluster-first route-second along with a greedy randomized adaptive search procedures (GRASP) meta-heuristic to resolve VRP. Simeonov et al. [14] developed a learning-dependent population VNS method to report the practical logistics issue motivated by a gas delivery firm in the United Kingdom. Even though VNS is a robust optimization model to withstand the complete local search, it has not been established with structured storage model.

Long-term memory infrastructure of EAs could enhance the limitations of VNS in a structured storage method. Traditional works are used in developing solutions according to EA as well as expert brilliance. A version of GA method is developed in [15] to optimize the routing distance according to practical domain of VRP. Sripriya et al. [16] presented a hybrid genetic search along with a diversity control under the application of GA to resolve VRP. The route that has minimum distance is interchanged by using a crossover and two mutation tools in this study. Zhang and Lee [17] proposed an enhancement of traditional ABC model to deal with VRP. As the EA-relied methods often require maximum duration to explore solutions, more number of techniques were used effectively like heuristic model as well as greedy techniques for solving VRP. An extended ACO integrates the semi-greedy heuristic NEH, which has been developed by Chen et al. [18] for VRP. In order to enhance, the ACO function boosts the fundamental ACO that eliminates local optimum and concatenates the adjacent searching model. Gupta and Saini [19] used the extended ACO, which comprises pheromone upgrading model and 2-opt model to increase the generalized route for VRP. Though VRP is assumed with practical limitations for logistics, the issues are regarding the mechanism of sharing few users. As people become busier, the logistics services to pick up any things from customers are considered to be more significant. Hence, managing two diverse kinds of logistics services are the severe problem. Several methods available in the literature are presented in [20-22].

The similarities of these techniques are used to combine the intelligence and knowledge from professionals at the time of iteration process to find optimal logistics solutions in an effective fashion. Mostly, such techniques are applied to explore optimized logistics solution when compared with heuristic approaches at same implementation time. But the storage as well as processing cost is a major challenge since it has massive data while the iterations are saved. The logistics industry does not provide a sufficient processing resource to operate an evolutionary model. Also, the purpose of existing works is to reduce the overall driving distance. It has limited the count of vehicles applied that influences total processing cost of logistics. These models are termed as first-delivery-last-pickup (FDLP). It is evident that the vehicle's space could be applied effectively when a courier is filled with products from users on a back way. A set of three issues have been resolved by the HFMPSO model as given below.

- A package has to be divided. In several cases, a vehicle is not capable of managing massive package count by a wider margin. It defines that the packages should be classified into few bunches for transport facility.
- Vehicle space control: When there are two types of logistics needs that are to be processed at the same time, the courier has to verify whether the vehicle has sufficient space to carry packages at any time and place as vehicle's ability is fine.

- Consuming quality and efficiency: Several real firms of logistics are desired to attain an optimized logistics solution.

2.3 PROPOSED METHOD

Figure 2.1 shows the overall system framework of the HFMPSO model. The proposed logistics platform is composed of few vehicles that have restricted capacity and a logistic graph contains a depot and few users are from urban area. At the primary stage, every delivery package has been gathered from a depot and each pickup package is placed at the nearby users. In HFMPSO, it is fixed with an upper bound of loading rate to select the number of packages that a vehicle can hold. HFMPSO is constrained with solution generation. It is mainly employed to explore the suitable logistics solution, which is comprised of three phases: package dividing, route scheduling, and package insertion. Hence, every package is classified as massive logistic path and a logistics solution is produced to transmit packets. As the logistics conditions as well as the needs vary in a rapid manner, it has been decided to attain the suitable logistics solution in an effective manner. On the other hand, it is evident that vehicle's space is applied productively when a courier carries the packages from clients on a back way; hence, the upper bound is named as loading rate ρ, which is developed to model the way of holding a package from a depot. The loading rate ρ is considered as 0. It refers to a higher vehicle potential in package delivery.

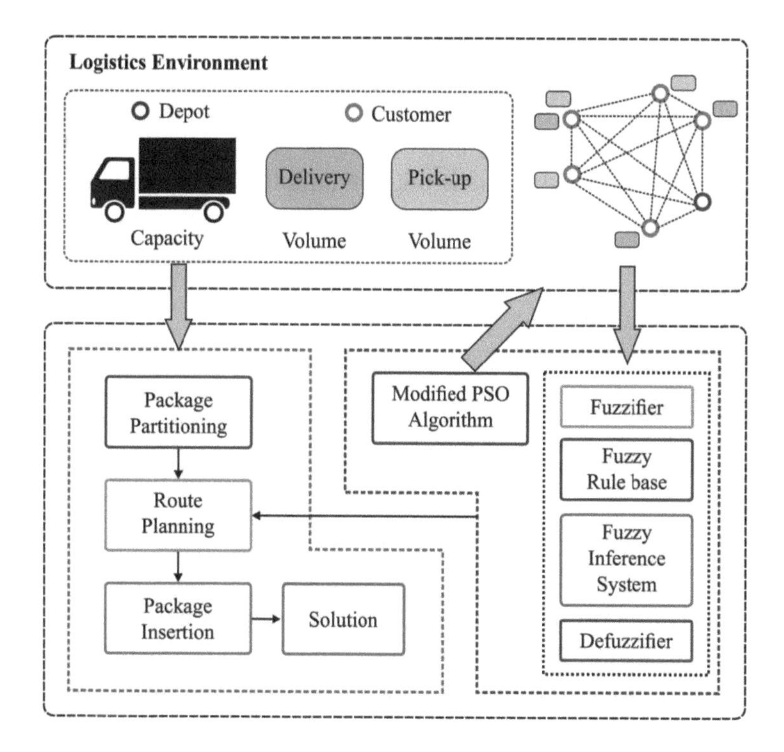

FIGURE 2.1 Overall architecture of HFMPSO algorithm.

2.3.1 Package Partitioning

Since there are massive packages gathered at depot periodically where the ability of a vehicle is minimum, it is essential to compute which packages have to be provided. Here, four principles are used for selecting the packages in the form of a batch that meets the vehicle's potential. The selection procedures are termed as near customer-first (NCF) as well as far-customer-first (FCF), which is based on the distance from a depot to nearby users. The major techniques of NCF and FCF are used in selecting packages which are nearby and far away from a depot. NCF would decide the p_2^d and p_3^d as two adjacent users c_3 and c_5. The residual ability is not enough to capture p_1^d. Therefore, the first and second batches are $\{p_2^d$ and $p_3^d\}$ and $\{p_1^d\}$, correspondingly. $\{p_1^d$ and $p_3^d\}$ and $\{p_2^d\}$ are two batches for *FCF*.

The division of NCF and FCF are similar to concentric circles since the distances are same in an identical batch. Then, it is optimal when packages filled with users placed in a nearby area are collected as a batch. Hence, near-NCF (NNCF) and near-FCF(NFCF) are developed to divide the package. The NNCF and NFCF search the anchor package p', which has the delivery locations closer to and farthest from a depot, respectively. The delivery place of packages that is nearby location of anchor package is provided with first preference. The nearest user to depot is c_3. p_2^d has been induced first to vehicle by NNCF and the rest are 5. Then, a nearby customer to deliver from c_3 is c_5. Hence, it is loaded with p^{d_3} into the vehicle. The overall loaded amount is actually same as the potential of a vehicle. Therefore, $\{p_2^d$ and $p_3^d\}$ and $\{p^{d_1}\}$ are declared as the first and second batches. In contrast, $\{p^{d_1}$ and $p_2^d\}$ and $\{p_3^d\}$ are two batches for NFCF.

2.3.2 Planning of Delivery Path Using HFMPSO Algorithm

The default nature of animals, birds, and fishes are the major evolution to model PSO technique. The collection of elements migrates by feasible space to find an optimal outcome. The position and velocity of each particle might be implied by a position as well as velocity vectors, along with an index value of *I*. P_{best} and G_{best} are assumed to be best position of every particle and between all rounds of components. The numerical presentation is expressed as given below:

New velocity (V):

$$V_I(T+1) = WV_I + \xi_1\left(P_{best} - X_I(T)\right) + \xi_2\left(G_{best} - X_I(T)\right), \quad I = 1, 2, \ldots, N; \; T = 1, 2, \ldots, d \tag{2.1}$$

New position (X):

$$X_I(T+1) = X_I(T) + V_I(\mathrm{T}+1) \tag{2.2}$$

where T shows the function of discrete time index along with a range of search space dimension (d). The acceleration coefficients are ξ_1 and ξ_2, thus the particle velocity can be measured by an inertia weight (W) to upgrade the position. The impact of optimal velocity as well as position of a particle could be managed under the application of ξ_1 and ξ_2. Here, equation (2.1) applies the velocity clamping model to control the process. To limit the velocity increment, band across the limit might be viable by the alterations of equation (2.1) as provided:

$$V_I(T+1)=\sigma\left[V_I(T)+\xi_1\left(P_{best}-X_I(T)\right)+\xi_2\left(G_{best}-X_I(T)\right)\right] \tag{2.3}$$

where

$$\sigma = 2\left(\left|2-\vartheta-\sqrt{\vartheta^2-4\vartheta}\right|\right)^{-1}, \quad \text{when } \vartheta=\xi_1+\xi_2>2 \tag{2.4}$$

The difference of equation (2.1) is functioned with the application of constriction factor method (CFM) that is represented by a variable σ.

The main goal of this HFMPSO model is to design a new shortest path technique by applying fuzzy rules to reduce the cost and time consumption to reach the desired result. It is extended by implementing HFMPSO to apply the restricted estimation along with some fuzzy rules. The organizations of indeterminate most reduced problem by assuming fuzzy path lengths have been estimated with transportation. Since a meta-heuristic approach is robust, PSO measurement is modified and used to manage the fuzzy shortest path technique. The dynamic operation is optimized to extend the function of regular system and release from an untimely merging PSO model. Here, the MPSO technique is applied to compute the FSPP.

The searching ability might be influenced by an encoding system of network inside a particle. In the developed model, particle position has been denoted in the form of priority vector to help the nodes build a shortest path. Basically, the path generation is an arbitrary format from source to destination node. It is assumed with unorganized nodes of route development in all rounds. However, the projected method uses the random path by first priority nodes to generate paths to attain the destination. Also, the ineffective path might be emerged as the invalid termination of path exists in an accurate destination.

The matching of input with output is feasible by a fuzzy interference system. Moderate convergences as well as local optimum termination, are assumed to be two major errors of PSO model. The improved velocity has been limited as the specific threshold value is lower. The velocity value as well as inertia weight are inversely proportional to one another, which influence the FF. A regular strategy for PSO model is depicted in Figure 2.2.

The FF level of definite particle is derived by an operation of τ, current best performance evolution (CBPE) from a fuzzy-based MPSO. Naturally, the good and bad FF values attained may be defined as α and β. Likewise, the normalized CBPE (NCBPE) could be determined by applying the given function:

$$NCBPE, \bar{\tau} = \frac{\tau-\alpha}{\beta-\alpha} \tag{2.5}$$

A and B are two parameters employed in this model to define the fuzzy function along with an assumption of $\bar{\tau}$ and Z as input and output rates.

$$A=\left|P_{\text{best}}-X\right| \text{ and } B=\left|G_{\text{best}}-X\right| \tag{2.6}$$

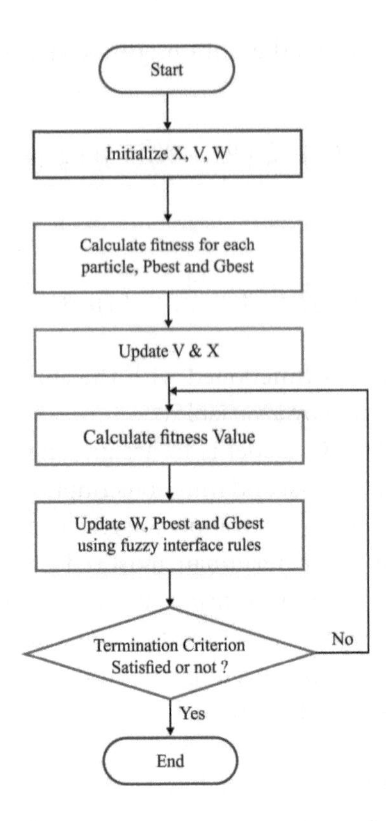

FIGURE 2.2 Algorithmic flow of PSO.

The distance between particle's current location and the local and global best is referred to as A and B, respectively. The A and B measures are allocated on the basis of size in a search space, which helps to select the rate of Z in a radius of $[0, 1]$. The procedures of fuzzy interference system can be applied to select the optimized result, as provided in Table 2.1. According to the three inputs Z, the simulation outcome would be chosen in a fuzzy model.

When the three inputs are less, then the adjacent FF is sufficient to produce minimum resultant value. Then, the searching phase is repeated to the value of global optimum, and the output measure is higher for every combination of input rules. Likewise, the particle is closer to local optimum by replacing the global rates for third rule. Similarly, MPSO has been applied to resolve the FSPP for irregular networks.

Here L indicates low, M refers to medium, and H denotes high.

2.3.3 Inserting Pickup Packages

This process involves pickup packages from user and buying them in return to depot. It is evident that pickup package has to be allocated delivery route, which may be nearer to pickup location. Hence, it is developed with allotment-based insertion (ABI) principles to declare every picking up package is for a nearby livery route. Also, it combines ABI mechanism with centroid distance $\left(ABI_{CD}\right)$, perpendicular distance $\left(ABI_{PD}\right)$, and additional distance $\left(ABI_{AD}\right)$ values. As given in Figure 2.4(a), o_1 implies a centroid of initial route, and p_1^p, p_2^p, and p_3^p come under the first delivery route as it is near o_1. Later, every

TABLE 2.1 Rules of Fuzzy Interference System

Input			Output
A	B	$\overline{\tau}$	Z
L	L	L	L
L	L	M	H
L	L	H	H
L	M	L	H
L	M	M	H
L	M	H	H
L	H	L	H
L	H	M	H
L	H	H	H
M	L	L	H
M	L	M	H
M	L	H	H
M	M	L	H
M	M	M	H
M	M	H	H
M	H	L	H
M	H	M	H
M	H	H	H
H	L	L	H
H	L	M	H
H	L	H	H
H	M	L	H
H	M	M	H
H	M	H	H
H	H	L	H
H	H	M	H
H	H	H	H

package is allocated to the nearby route segment. But the additional routing cost is used to pick up p_3^p from c_5 that is lower when compared with c_1 and even p_3^p is nearer to o_1 than o_2. In order to explain the scenario, the pickup packages have been declared on the basis of perpendicular distance for ABI_{PD} procedure. As depicted in Figure 2.4(b), p^{p_3} is adjacent to the second route. Figure 2.4(c) showcases the ABI_{AD}, as the included distance is $\mu(c_1, c_2)+\mu(c_2, c_3)-\mu(c_1, c_3)$, when c_2 is added among c_1 and c_3.

The result attained from insertion process under the application of ABI_{PD} and ABI_{AD} is varied to apply ABI_{CD} and the routing expense has been limited.

2.4 EXPERIMENTAL VALIDATION

To estimate the quality of a route and processing efficiency of HFMPSO algorithm, various types of experiments were carried out. The practical logistics data have been gathered from a logistic industry in Chennai as well as the collection of two cities named Pondicherry and Kanchipuram are selected for experimental dataset.

TABLE 2.2 Logistic Cost Analysis of Different Models under Varying Number of Packages (Pondicherry)

Number of Packages	FDLP	ILSP	Proposed
150	4590	5140	7700
200	4830	5240	7890
250	5030	5600	8000
300	5182	6000	8190
350	5358	5950	8310

2.4.1 Performance Analysis under Varying Package Count

Table 2.2 and Figure 2.3 illustrate the logistic cost prediction of different methods in a varying number of packages between Chennai and Pondicherry. It is depicted that the proposed method obtains higher logistic cost when compared with alternative techniques. With respect to total of 150 packages, it is implied that the presented system attains greater logistic cost of 7700, while the FDLP and ILSP approaches reached a minimum logistic cost of 4590 and 5140, respectively. Simultaneously, by using the package value of 200, the FDLP model has provided a lower logistic cost of 4830 and the ILSP method achieved a better logistic cost of 5240. However, the projected system displays slightly better outcome by reaching a higher logistic cost of 7890. Likewise, by applying the package number of 250, the deployed method shows a maximum logistic value of 8000 and FDLP and ILSP frameworks attained a less logistic cost of 5030 and 5600, respectively. In the same way, using the package number of 300, it is demonstrated that the proposed system has shown productive results by reaching higher logistic cost of 8190, while the FDLP and ILSP technologies accomplish a minimum logistic cost of 5182 and 6000, respectively.

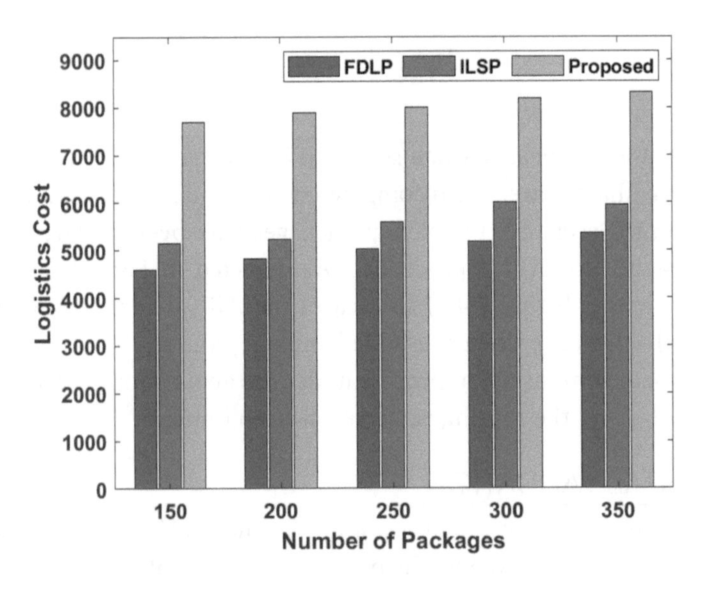

FIGURE 2.3 Logistic cost analysis of different models under varying number of packages (Pondicherry).

TABLE 2.3 Logistic Cost Analysis of Different Models under Varying Number of Packages (Kanchipuram)

Number of Packages	FDLP	ILSP	Proposed
150	4748	5026	8356.5
200	5047	5737	8552
250	5300	5783	8713
300	5541	6185	8920
350	5656	6335	9046.5

Table 2.3 and Figure 2.4 show the logistic cost analysis of diverse models under a varying number of packages between Chennai and Kanchipuram. It is shown that the HFMPSO model attains maximum logistic cost over the other models. Under a total of 150 packages, it is shown that the HFMPSO model achieves a maximum logistic cost of 8356, whereas the FDLP and ILSP models obtained a lower logistic cost of 4748 and 5026, respectively. At the same time, under the package count of 200, the FDLP model has offered a minimum logistic cost of 5047 and the ILSP model has reached to a slightly higher logistic cost of 5737. But the HFMPSO model shows better results by achieving a maximum logistic cost of 8552. Similarly, under the package count of 250, the HFMPSO model exhibits a higher logistic count of 8713, whereas the FDLP and ILSP models reached to a minimum logistic cost of 5300 and 5783, respectively. Likewise, under the package count of 300, it is exhibited that the HFMPSO model has demonstrated effective results by attaining a maximum logistic cost of 8920, whereas the FDLP and ILSP models obtained a lower logistic cost of 5541 and 6185, respectively.

2.4.2 Performance Analysis under Varying Vehicle Capacities

Table 2.4 and Figure 2.5 illustrate the logistic cost prediction of different methods in a varying number of vehicles between Chennai and Pondicherry. It is depicted that the proposed

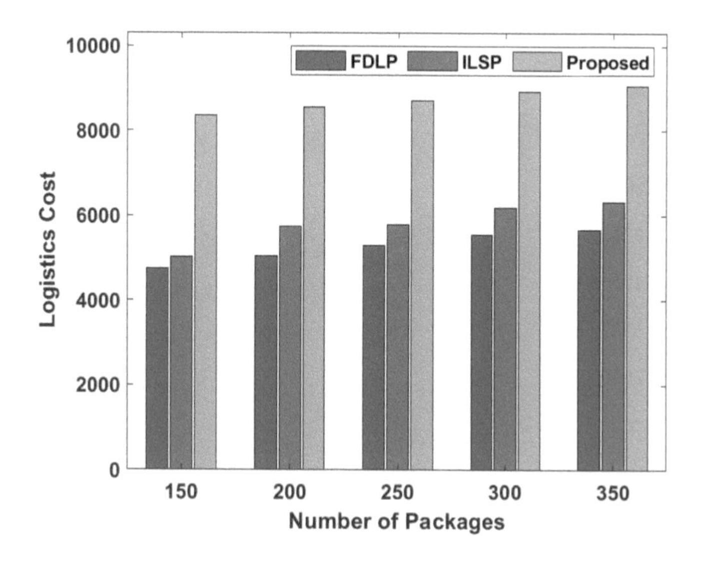

FIGURE 2.4 Logistic cost analysis of different models under varying number of packages (Kanchipuram).

TABLE 2.4 Logistic Cost Analysis of Different Models under Varying Vehicle Capacities (Pondicherry)

Number of Packages	FDLP	ILSP	Proposed
150	4890	5430	7780
200	4820	5380	7800
250	4830	5360	7900
300	4810	5370	7938
350	4810	5370	8030

method obtains higher logistic cost when compared with alternative techniques. With respect to total of 150 vehicles, it is implied that the presented system attains greater logistic cost of 7780, while the FDLP and ILSP approaches reached a minimum logistic cost of 4890 and 5430, respectively. Simultaneously, by using the vehicle count of 200, the FDLP model has provided a lower logistic cost of 4820 and the ILSP method achieved a better logistic cost of 5380. However, the projected system displays slightly better outcome by reaching a higher logistic cost of 7800. Likewise, by applying the vehicle capacity of 250, the deployed method shows a maximum logistic value of 7900 and FDLP and ILSP frameworks attained a less logistic cost of 4830 and 5360, respectively. In the same way, using the vehicle capacity of 300, it is demonstrated that the proposed system has shown productive results by reaching higher logistic cost of 7938, while the FDLP and ILSP technologies accomplish a minimum logistic cost of 4810 and 5370, respectively. Finally, under the vehicle capacity of 350, it is shown that the HFMPSO model reaches a maximum logistic cost of 8030, whereas the FDLP and ILSP models have attained a lower logistic cost of 4810 and 5370, respectively.

Table 2.5 and Figure 2.6 show the logistic cost analysis of diverse models under a varying number of vehicle capacity between Chennai and Kanchipuram. It is shown that the HFMPSO model attains maximum logistic cost over the other models. Under a total of

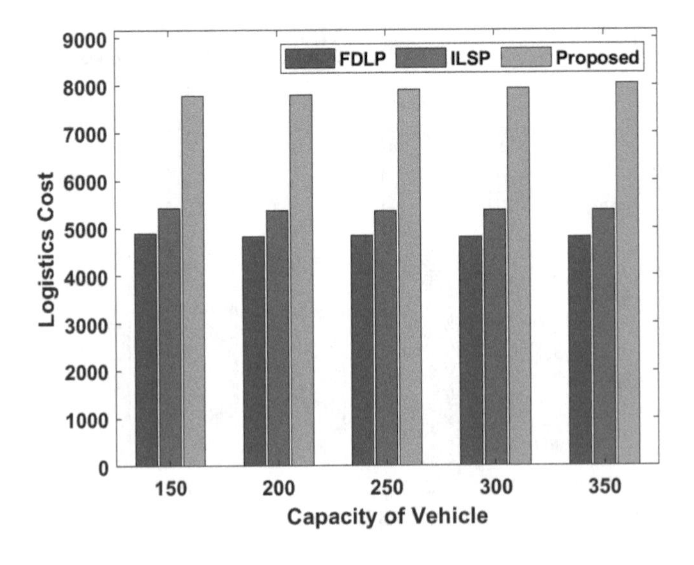

FIGURE 2.5 Logistic cost analysis of different models under varying vehicle capacities (Pondicherry).

TABLE 2.5 Logistic Cost Analysis of Different Models under Varying Vehicle Capacities (Kanchipuram)

Varying Vehicle Capacity	FDLP	ILSP	Proposed
150	4870	5000	7778
200	4810	5550	7810
250	4820	5390	7890
300	4820	5378	7920
350	4810	5360	7400

150 vehicle capacities, it is shown that the HFMPSO model achieves a maximum logistic cost of 7778, whereas the FDLP and ILSP models obtain a lower logistic cost of 4870 and 5000, respectively. At the same time, under the vehicle capacity of 200, the FDLP model has offered a minimum logistic cost of 4810 and the ILSP model has reached a slightly higher logistic cost of 5550. But, the HFMPSO model shows better results by achieving a maximum logistic cost of 7810. Similarly, under the vehicle capacity of 250, the HFMPSO model exhibits a higher logistic count of 7890, whereas the FDLP and ILSP models reached a minimum logistic cost of 4820 and 5390, respectively. Likewise, under the vehicle capacity of 300, it is exhibited that the HFMPSO model has demonstrated effective results by attaining a maximum logistic cost of 7920, whereas the FDLP and ILSP models obtain a lower logistic cost of 4820 and 5378, respectively. Besides, under the vehicle capacity of 350, the HFMPSO model has demonstrated effective results by attaining a maximum logistic cost of 7400, whereas the FDLP and ILSP models obtained a lower logistic cost of 4810 and 5360, respectively.

2.4.3 Computation Time (CT) analysis

Figure 2.7 shows the CT analysis of diverse models under a varying number of packages from a transmit between Chennai and Pondicherry. It is shown that the HFMPSO model

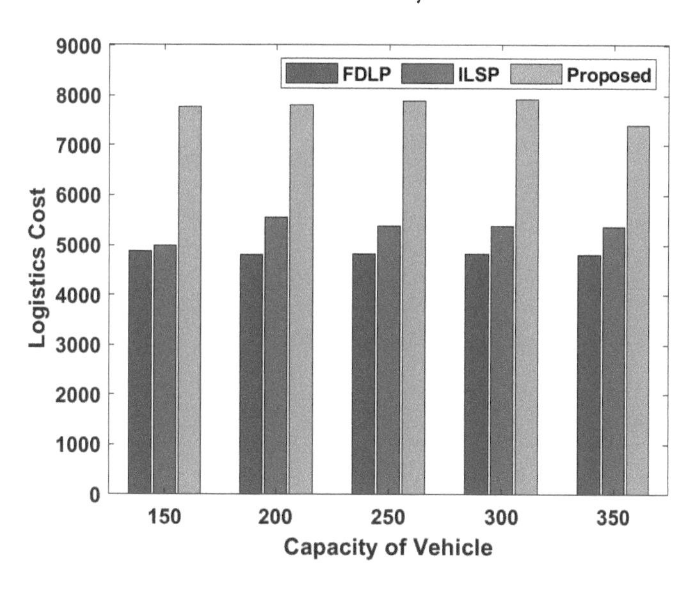

FIGURE 2.6 Logistic cost analysis of different models under varying vehicle capacities (Kanchipuram).

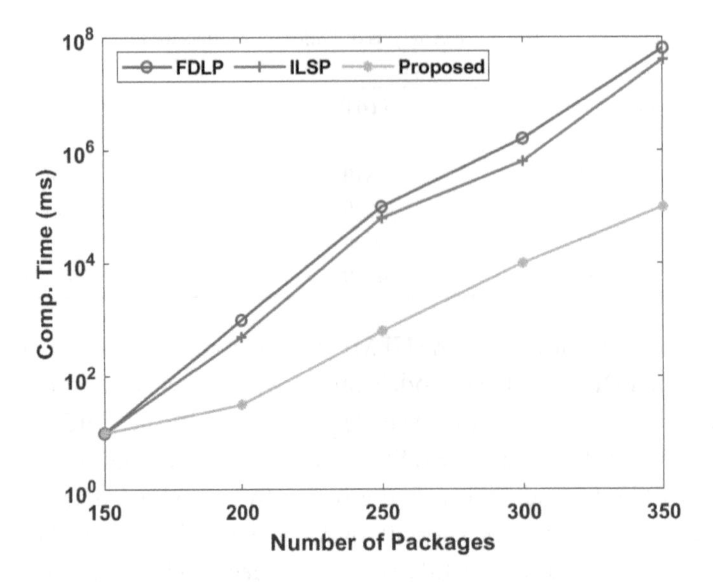

FIGURE 2.7 CT analysis under varying package count (Pondicherry).

requires minimum CT over the compared methods. For instance, under the presence of 150 packages, the HFMPSO model requires a minimum CT of 80 ms, whereas the existing FDLP and ILSP models need a maximum CT of 3000 and 3162 ms, respectively. On the other side, under the maximum package count of 300, it is noted that the HFMPSO model offers a least CT of 10^5 ms, whereas the existing FLDP and ILSP models have offered a maximum CT of 10^8 ms.

Figure 2.8 examines the CT offered by diverse models under a varying number of packages from a transmit between Chennai and Kanchipuram. It is evident that the HFMPSO

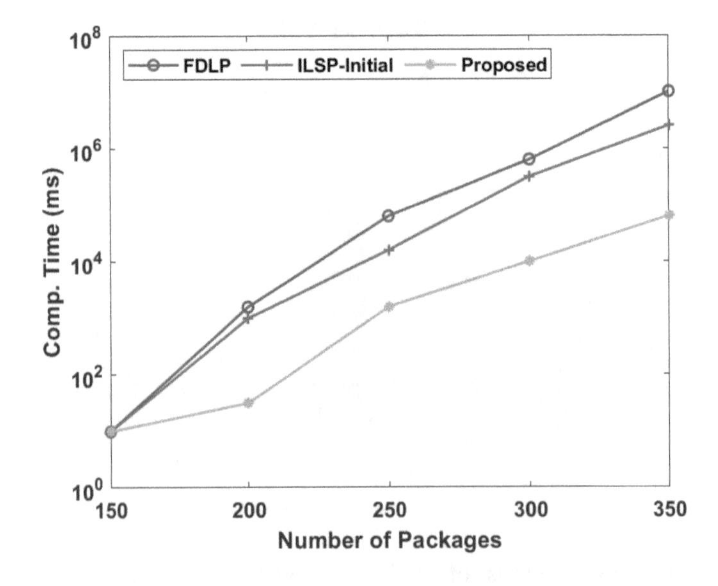

FIGURE 2.8 CT analysis under varying package count (Kanchipuram).

model requires minimum CT over the compared methods. For instance, under the presence of 150 packages, the HFMPSO model requires a minimum CT of 70 ms, whereas the existing FDLP and ILSP models need a maximum CT of 3100 and 3132 ms, respectively. On the other side, under the maximum package count of 300, it is noted that the HFMPSO model offers a least CT of around 32,000 ms, whereas the existing FLDP and ILSP models have offered a maximum CT of 10^7 and 10^6 ms, respectively. The above-mentioned detailed results analysis showcased the betterment of the HMFPSO model over the compared methods under diverse aspects.

2.5 CONCLUSION

This chapter has introduced a new intelligent logistic transportation planning model called HFMPSO algorithm. The proposed model initially used IoT-based barcode reader to access the details of the package. The proposed transportation planning algorithm involves three main phases: partitioning of packages, route planning using HFMPSO algorithm, and package insertion. At the primary stage, every delivery package has been gathered from a depot and each pickup package is placed at the nearby users. In HFMPSO, it is fixed with an upper bound of loading rate to select the number of packages that a vehicle can hold. HFMPSO is constrained with solution generation. The detailed experimental outcome clearly exhibited the superior performance of the proposed model over the compared methods.

ACKNOWLEDGMENT

The author gratefully acknowledges Ho Chi Minh City Open University, Vietnam for supporting this research.

REFERENCES

1. Han Z., Li X., Zhang T., Chen S. (2015). Research on the influence mechanism from changing logistics demand of manufacturing industry to the logistics industry based on the big data analysis. *Logist Engineering Management*, 5, 23–26.
2. Kawa A. (2017). Fulfillment service in e-commerce logistics. *LogForum*, 13, 429–438.
3. Wang Y., Feng L., Chang H., Wu M. (2017). Research on the impact of big data on logistics. *MATEC Web of Conferences*, 100, 02015.
4. Zhong R. Y., Newman S. T., Huang G. Q., Lan S. (2016). Big data for supply chain management in the service and manufacturing sectors: challenges, opportunities, and future perspectives. *Computers & Industrial Engineering*, 101, 572–591.
5. Kataria M., Mittal M. P. (2014). Big data: a review. *International Journal of Computer Science and Mobile Computing*, 3, 106–110.
6. Aliresearch. (2017). Big Data Development Report on China Smart Logistics. http://i.aliresearch.com/img/20170328/20170328173815.pdf. Accessed 18 June 2018.
7. Hofmann E., Rüsch M. (2017). Industry 4.0 and the current status as well as future prospects on logistics. *Computers in Industry*, 89, 23–34.
8. Chami Z. Al, Manier H., Manier M.-A., Fitouri C. (2017). A hybrid genetic algorithm to solve a multi-objective pickup and delivery problem. *IFAC: PapersOnLine*, 50, 14656–14661.
9. Marinaki M., Marinakis Y. (2016). A glowworm swarm optimization algorithm for the vehicle routing problem with stochastic demands. *Expert Systems with Applications*, 46, 145–163.
10. Osman I. H. (1993). Metastrategy simulated annealing and Tabu search algorithms for the vehicle routing problem. *Annals of Operations Research*, 41(4), 421–451.

11. Huber S., Geiger M. J. (2014). Swap Body Vehicle Routing Problem: A Heuristic Solution Approach. In: *International Conference on Computational Logistics*, 24–26 September 2014, Chile, 16–30.
12. Sze J. F., Salhi S., Wassan N. (2016). A hybridisation of adaptive variable neighbourhood search and large neighbourhood search: application to the vehicle routing problem. *Expert Systems with Applications*, 65, 383–397.
13. Miranda-Bront J. J., Curcio B., Méndez-Díaz I., Montero A., Pousa F., Zabala P. (2017). A cluster-first route-second approach for the swap body vehicle routing problem. *Annals of Operations Research*, 253(2), 935–956.
14. Simeonov L., Wassan N., Salhi S., Nagy G. (2018). The heterogeneous fleet vehicle routing problem with light loads and overtime: formulation and population variable neighbourhood search with adaptive memory. *Expert Systems with Applications*, 114, 183–195.
15. Tunjongsirigul B., Pongchairerks P. (2010). A Genetic Algorithm for a Vehicle Routing Problem on a Real Application of Bakery Delivery. In: *The 2nd International Conference on Electronic Computer Technology*, 7–10 May 2010, Malaysia, 214–217.
16. Sripriya J., Ramalingam A., Rajeswari K. (2015). A Hybrid Genetic Algorithm for Vehicle Routing Problem with Time Windows. In: *International Conference on Innovations in Information, Embedded and Communication Systems*, 19–20 March 2015, India.
17. Zhang S. Z., Lee C. K. M. (2015). An Improved Artificial Bee Colony Algorithm for the Capacitated Vehicle Routing Problem. In: *IEEE International Conference on Systems, Man, and Cybernetics*, 9–12 October 2015, China, 2124–2128.
18. Chen R.-M., Hsieh F.-R., Wu D.-S. (2012). Heuristics Based Ant Colony Optimization for Vehicle Routing Problem. In: *The 7th IEEE Conference on Industrial Electronics and Applications*, 18–20 July 2012, Singapore, 1039–1043.
19. Gupta A., Saini S. (2017). An Enhanced Ant Colony Optimization Algorithm for Vehicle Routing Problem with Time Windows. In: *The 9th International Conference on Advanced Computing*, 14–16 December 2017, India, 267–274.
20. Devaraj A. F. S., Elhoseny M., Dhanasekaran S., LaxmiLydia E., Shankar K. (2020). Hybridization of firefly and improved multi-objective particle swarm optimization algorithm for energy efficient load balancing in cloud computing environments. *Journal of Parallel and Distributed Computing*, 142, 36–45.
21. Elhoseny M., Shankar K., Lakshmanaprabu S. K., Maseleno A., Arunkumar N. (2018). Hybrid optimization with cryptography encryption for medical image security in Internet of Things. *Neural Computing and Applications*, 32, 10979–10993. https://doi.org/10.1007/s00521-018-3801-x
22. Lakshmanaprabu S. K., Mohanty S. N., Rani S. S., Krishnamoorthy S., Uthayakumar J., Shankar K. (2019). Online clinical decision support system using optimal deep neural networks. *Applied Soft Computing*, 81, 1–10.

CHAPTER 3

Butterfly Optimization-Based Feature Selection with Gradient Boosting Tree for Big Data Analytics in Social Internet of Things

Irina V. Pustokhina[1] and Denis A. Pustokhin[2]

[1]*Department of Entrepreneurship and Logistics, Plekhanov Russian University of Economics, Moscow, Russia*

[2]*Department of Logistics, State University of Management, Moscow, Russia*

CONTENTS

3.1 INTRODUCTION

The fundamental objective of Fourth Industrial Revolution, deployment and extensive application of Internet of Things (IoT) system, resulted in the enhancement of connectivity among humans, human to object, and object to device, reinforcing an emergence of hyper-connected public community [1]. Based on the statement, it is desired that a fresh metric of fusion services appear after the interconnecting massive smart tools. Besides, in hyper-connected society, intellectualized objects will not apprehend the human situation, be aware of their needs (Context Aware), rather independently provide and recommend the better solution; however, it supports in taking control measures as the human requirements [2]. The domain that helps to develop creative and effective service at a rapid pace (new measure) perform the extension of previous service (Tapped Value), which is meant to be Social IoT (SIoT) model.

Based on Ref. [3], the SIoT is referred to as the IoT in which the objects are applicable to develop social combinations with each other independently. In addition, SIoT concept shows the environment that enables people as well as objects to communicate inside a social structure of associations [4]. SIoT platform supports the organization by accomplishing diverse and massive amount of data, which has been collected named as Big Data that is said to be a smart service by processing a better interference [5]. It refers to the dynamic format of mutual data along with the objective of required service that exceeds a distribution and reading level of a text, image, and so forth in previous social media. Furthermore, the range of Social Network Service has been upgraded from single targeted to corporation targeted, which is often with IoT, and finally activates the business functional collaboration [6].

SIoT is simulated to be one of the well-known concepts for massive state-of-the-art, rapidly developing models, and some of them are IP-enabled embedded devices as well as smart objects, short- and long-range communication models, data accumulation, examination, computation, and visualization devices from big market giants that have various benefits in network direction, reliability, estimation of objects, security, service composition, object identification, behavior categorization, as well as detection. According to the parameters of SIoT, different types of studies were developed about developing a method for SIoT service environment and significance of the domains applied. In addition, the related works are carried out [7] such as performing the interference, which depends upon a research in data gathering process inside the SIoT platform as well as the gathered big data, and research to maintain the efficiency of data and security.

Building the IoT-based models and corresponding solutions is considered to be the major challenging task. Hence, IoT deals with the persistent accumulation as well as data distribution to a general objective [8]. In IoT, data means the parameter values like variables or integer measures; and it depicts some specialized conditions that are accomplished [9, 10]. IoT networks enable little functionality by using a previous interface. Developers have highly concentrated in finding threats while identifying and combining the data inside IoT [11]. Thus, SIoT is said to be a massive social network that connects people to people, people to objects, and objects to objects [12]. Therefore, developing the opportunities provide major challenging issues to data computation for the purpose of enhancing data collection, noise elimination, storage, as well as to perform actual analytics [13]. Furthermore, the big data contains diverse standards and devices by relational database vendors, and it is also applied for data collection and data analysis [14]. Here, "big data and SIoT are the exact depiction of social systems and IoT to simplify human development" [15]. Different types of feature selection (FS) model have been projected and divided into two classes, namely, filter and wrapper models. Initially, filter-enabled method performs the filtration task that has to be computed in prior to classify the data because of the random application of classification models [16].

To resolve the big data issues existing in SIoT, this chapter devises a new big data analytics method in SIoT using butterfly optimization-based feature selection with a gradient boosting tree (GBT) technique called BOAFS-GBT. The proposed BOAFS-GBT model primarily performs feature selection using the BOAFS model, which selects a constructive collection of features from the big data. Thereafter, the GBT model is used for the classification of the feature-reduced data into several classes. In addition, big data Hadoop framework has been employed for big data processing. The outcome of the BOAFS-GBT model has been validated against three datasets under diverse aspects.

3.2 RELATED WORKS

SIoTs alludes to rapid development of related protests and humans are able to collect and interchange the data with the help of in-built sensors presented by Hasan and Al-Turjman [17]. Authors have presented a nature-based particle multi-swarm enhancement (PMSO) that navigates to develop, reform, and choose disjoint ways that carry on the disappointment at the time of satisfying quality of service (QoS) attributes. The MS approach empowers selecting the best directions in deciding multipath directing, and simultaneously the trading messages from each position in the model. The final outcome represents that the method that has applied the qualities of optimal data is one of the substantial models for motivations behind improving the PMSO implementation. High-performance computing (HPC) solution is referred to be a major problem and developed by Ahmad et al. [18]. It is applied with a technique that selects artificial bee colony (ABC). Moreover, a Kalman filter (KF) has been employed as a portion of Hadoop biological system, which is applied for noise elimination. The four-level engineering is used for removal of superfluous data, and the data obtained are examined with the help of presented Hadoop-based ABC model. For validating the efficiency of newly deployed approach in framework engineering, the newly deployed Hadoop and Map Reduce with ABC computation are used. ABC estimation is applied for selecting the better highlights; even though Map Reduce is applied in all applications, it consumes massive quantity of data. The IoT is overloaded by maximum objects with numerous communications and facilities.

Mardini et al. [19] suggested the SIoT, in which all questions in IoT can apply the companions' or friends-of-friends' connections that seek for specific management. Generally, it is common strategy for all objects that are essential to resolve the substantial companions. It intends to resolve the problem of link calculation of companions and scrutinizes five models in this work. It is presented with a link estimation principle under the application of genetic algorithm (GA) for better result. The results depicted in examined applications modify some attributes. Hence, some of the complexities are as follows:

- IoT devices are used for interfacing with one another effectively, and finally developing Big Data does not offer good reduction [20–22].
- The SIoT big data reduces the vital application with SIoT while interacting data over the Internet.
- Latest data analysis and modern technologies would offer the SIoT with optimization as well as classifier model.

3.3 THE PROPOSED METHOD

Once the data are created by SIoT devices, several transformations occur by utilization of transmit engines such as moving, cleaning, splitting, and merging. Thereafter, the information is saved in several methods such as cloud or various databases. Afterward, the proposed BOAFS-GBT model primarily performs feature selection using BOAFS model, which selects a constructive collection of features from the big data. Next, the GBT model is used for the classification of the feature-reduced data into several classes. In addition, big data Hadoop framework has been employed for big data processing. These processes are illustrated in Figure 3.1.

3.3.1 Hadoop Ecosystem

In order to manage big data, Hadoop ecosystem and corresponding units are applied. In a common platform, Hadoop is defined as a type of open-source structure that activates the

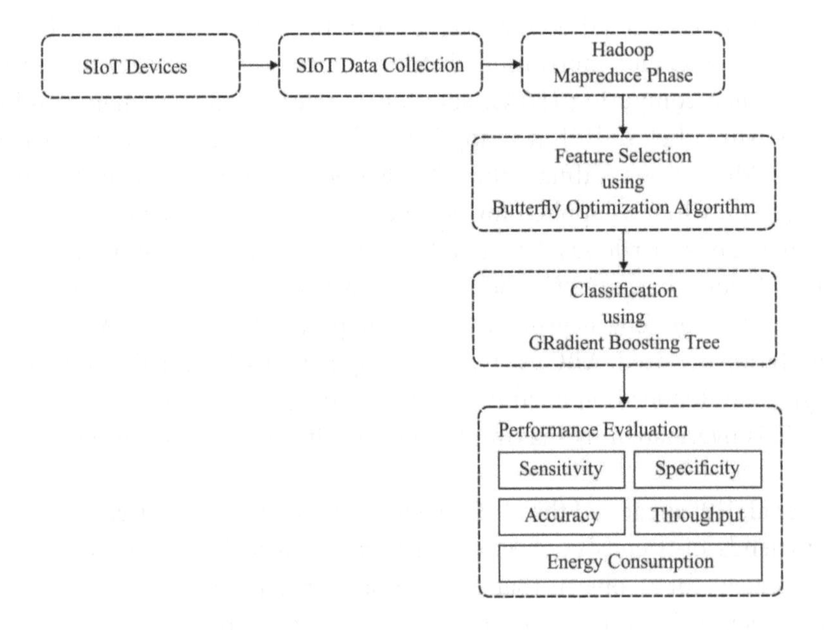

FIGURE 3.1 Block diagram of proposed method.

stakeholders for saving and computes big data over computer clusters using simple programming techniques. A massive node is fixed in a single server, and it contains improved reliability as well as fault tolerance. Key aspects of Hadoop are Map Reduce, Hadoop Distributed File System (HDFS), and Hadoop YARN.

3.3.1.1 Hadoop Distributed File System

According to Google File System (GFS), the HDFS has been labeled. It can be referred to as master/slave model, where the master has massive data nodes named as actual data while the alternative node is termed as metadata.

3.3.1.2 Hadoop Map Reduce

In order to offer massive scalability even under numerous Hadoop clusters, Hadoop Map Reduce has been applied named as programming model at Apache Hadoop heart. For computing enormous data over maximum clusters, Map Reduce can be applied. There are two vital stages in Map Reduce job processing like Reduce and Map stage. Each stage is composed of pairs such as key-value as input as well as output; in particular, in file scheme, output as well as input of a process is stored.

This approach computes task scheduling and management and reimplements the failed execution. The Map Reduce is composed of a slave node manager and single master resource manager to cluster nodes.

3.3.1.3 Hadoop YARN

It can be defined as a model applied to cluster management. Using the accomplished knowledge of initial Hadoop generation, it can be referred to as a second Hadoop generation that is meant to be a major objective. Among the Hadoop clusters, it provides security, reliable task, and data maintenance devices; YARN is applied as major approach and resource manager. For the big data management, additional devices and elements are deployed over the Hadoop formation. Figure 3.2 depicts the environment of Hadoop that are applied for managing big data proficiently.

3.3.1.4 Map Reduce Implementation

In processing a Map Reduce approach, MRODC model has been applied for enhancing the classification scalability and efficiency. Some of the factors comprised by MRODC models are as follows:

- Based on the N-gram, calculate all sentence polarity score.
- Based on the polarity measure, data classification is performed.
- Based on the classified data, determine new words and calculate the term frequency.

Under the application of diverse text mining models, first data from HDFS undergoes preprocessing. By using Map function, the concurrent iteration is processed named as Combiner purpose and diminish purpose, respectively.

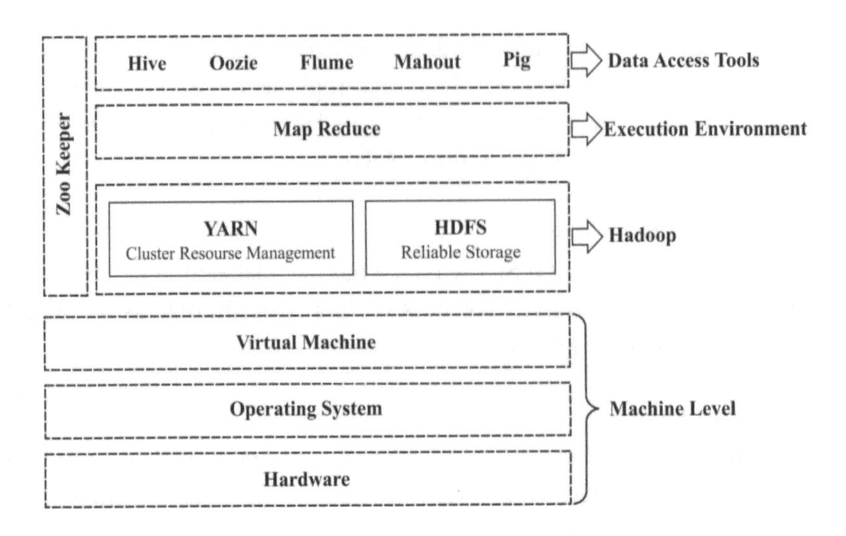

FIGURE 3.2 Hadoop framework.

3.3.1.4.1 Map Phase The objective of Map approves all lines sequentially as diverse pairs of key-value, which developed the Map function input. According to the developed corpus, Map function determines all data object values, and based on the several grams, the scores are evaluated and mapping function result is forwarded to Combiner function.

3.3.1.4.2 Combine Phase A complete data objects are retrieved from Combiner function, and Map function as well as data classification is carried out according to identical class. Next, it concatenates the complete data through same class values, in identical class, which saves the instance counts and Reducer function, and the final result is transmitted.

3.3.1.4.3 Reduce Phase In different classes, Reduce function obtains all data from these phases, which is the result of Combiner function. Thereafter, the summations of all information in diverse class labels are determined and a final result is secured in HDFS with class labels as well as consecutive iteration is invoked.

3.3.2 BOA-Based FS Process

In this section, the BOAFS algorithm gets executed to select the useful set of features. Nature-based metaheuristic approach has attained maximum attention from diverse applications in last decades. Butterfly optimization algorithm comes under the class of bioinspired model that is defined as a subclass of nature-based metaheuristic method. Basically, BOA is evolved from the food foraging nature of butterflies, and it is applied as searching agents that carry out the optimization in BOA. By default, butterflies comprise sense receptors that are used for sensing the smell of food or flowers. The sense receptors are named as chemoreceptors that have been distributed in butterfly's overall body.

Additionally, the butterfly can generate smell with some power. It is associated with the fitness of a butterfly determined under the application of objective function of a problem. It represents that if a butterfly shifts from one place to another in a search space, then the fitness will be changed.

The smell that is generated by a butterfly can be sensed by various other butterflies in the surroundings and aggregate social learning mechanism is followed. If the butterfly smells a fragrance from optimal butterfly in search space, then it develops a stride to good butterfly, which is termed as global search phase of BOA. Second, if a butterfly is unable to predict the fragrance of another butterfly in search space, which can develop arbitrary strides, it is termed as local search phase. In BOA, the scent is referred to as a function of external intensity of stimulus in the following:

$$pf_i = cI^a \tag{3.1}$$

where pf_i implies the received magnitude of scent, which determines the intensity of a fragrance of ith butterfly that is perceived by alternate butterflies, c depicts the sensory modality, I represents the stimulus intensity, and a shows the power exponent on the basis of modality, which consumes for different degree of absorption. In BOA, the artificial butterfly contains a position vector that is extended at the time of optimization process with the help of

$$x_i^{t+1} = x_i^t + F_i^{t+1} \tag{3.2}$$

where x_i^t denotes the solution vector x_i for ith butterfly with iteration value t and F_i showcases the scent applied by x_ith butterfly to upgrade the location at the time of iterations. Moreover, there are two major phases: global search phase and local search phase. Initially, a butterfly moves forward in a direction toward fittest butterfly or solution g^* that is represented by

$$F_i^{t+1} = \left(r^2 \times g^* - x_i^t\right) \times pf_i \tag{3.3}$$

Here g^* implies the recent best solution that is identified in new iteration. The obtained fragrance of ith butterfly is demonstrated by pf_i and r illustrates an arbitrary value from [0,1]. Local search phase is defined as follows:

$$F_i^{t+1} = \left(r^2 \times x_j^t - x_k^t\right) \times pf_i \tag{3.4}$$

where x_j^t and x_k^t are jth and kth butterflies from a solution space. When x_j^t and x_k^t come under similar population and r denotes a uniform random value from $[0,1]$ then equation (3.4) is referred to as a local random walk. A switch probability p has been applied in BOA to switch between common global search and intensive local search.

3.3.3 GBT-Based Classification

Breiman developed the bagging mechanism, which refers to the random sampling model for the purpose of training classification models. The classifiers are organized into a single group and provide maximum accuracy. In contrast to bagging with respect to sampling, boosting offers a weight for observation and alters the weight once the classifier training is completed. The weight of inaccurate classification has been improved, and the weight that is correctly classified observation is limited. The observations with changed weights are applied for training the upcoming classifier. Consequently, various classifiers undergo amalgamation uniformly. Friedman presented the GB approach. It is defined as a step-by-step model that concentrates on gradient reduction of loss function in predefined approaches.

The loss function is interpreted as a degree of error by newly developed approach. In general, if the loss function is maximum, then the method is more effective. The main objective is to reduce the loss function and erroneous rate, the optimal method is to reduce the loss function in gradient direction. Equation (3.5) shows the GB method in the following:

$$F(x;P) = F\left(x;\{\beta_{\mathrm{m}}, \alpha_{\mathrm{m}}\}_1^M\right) = \sum_{\mathrm{m}=1}^{M} \beta_{\mathrm{m}} h(x;\alpha_{\mathrm{m}}) \tag{3.5}$$

The $F(x;P)$ implies the χ function with p parameters and prediction function. Boosting is required for stacking several approaches. Then, β represents the node's weight and α is a parameter. It is feasible to enhance the prediction function F by optimizing $\{\beta, \alpha\}$. P implies the parameter of a method, and $\varphi(P)$ denotes a likelihood function of P. Equation (3.6) represents the loss function $F(x;P)$.

$$\varphi(P) = E_{y,x} L(y, F(x;P)) \tag{3.6}$$

When $m-1$, a number of methods are retrieved, the development of mth model could be accomplished with initial derivative to find the direction for loss function, where g_{m} is presented in equation (3.7):

$$g_m = \{g_{jm}\} = \left\{ \left[\frac{\partial \varphi(P)}{\partial p_j} \right]_{p=p_{m-1}} \right\} \tag{3.7}$$

Then, it is the deployment of gradient direction for likelihood function P for the proposed approach. Here, ρ_m showcases the distance of gradients as expressed in equation (3.8):

$$\rho_m = (lrg \min \varphi(p_{m-1} - \rho_m g_m) \tag{3.8}$$

At last, $f_m(x)$ function for mth model is retrieved, as shown in equation (3.9):

$$f_m(x) = -\rho_m g_m(x) \tag{3.9}$$

3.4 EXPERIMENTAL ANALYSIS

Table 3.1 provides a detailed dataset description of the presented model against three datasets, respectively. The GPS trajectories dataset includes a total of 163 instances under two class labels. Besides, the indoor user movement prediction dataset comprises a set of 13,197 instances, whereas 527 instances are present in water treatment plant.

The performance measures are used to examine throughput, energy consumption, sensitivity, specificity, and accuracy.

3.4.1 FS Results Analysis

Table 3.2 provides the analysis of the FS process of the BOAFS-GBT model in terms of number of features and optimized features.

On the applied dataset, the BOAFS-GBT model has chosen a set of 6 features out of 15 features from GPS trajectories dataset, 2 features from a set of indoor user movement prediction dataset, and 16 features from water treatment plant dataset.

3.4.2 Classification Results Analysis

Table 3.3 and Figures 3.3–3.5 perform the analysis of the classifier models of the proposed BOAFS-GBT model in terms of distinct performance measures. On the applied GPS trajectories dataset, the NBTree model has attained better results with the minimum sensitivity of 87.09%, specificity of 87.23%, and accuracy of 88.36%. At the same time, the MLP model has shown extraordinary results by achieving slightly higher sensitivity of 92.40%, specificity of 89.10%, and accuracy of 90.09%. Likewise, the SVM model has shown somewhat

TABLE 3.1 Dataset Description

Dataset	No. of Instances	Number of Classes
GPS trajectories	163	2
Indoor user movement prediction	13,197	2
Water treatment plant	527	2

TABLE 3.2 Results Analysis of Feature Selection

Dataset	No. of Features	Optimized Features
GPS trajectories	15	6
Indoor user movement prediction	4	2
Water treatment plant	38	16

TABLE 3.3 Results Analysis of Classification Performance

	GPS Trajectories			Movement Prediction			Water Treatment		
Methods	Sensitivity	Specificity	Accuracy	Sensitivity	Specificity	Accuracy	Sensitivity	Specificity	Accuracy
BOAFS-GBT	96.78	97.10	97.88	95.83	96.92	96.87	94.20	94.82	94.88
GBT	95.88	96.23	95.89	93.58	92.64	92.65	92.49	90.79	90.63
SVM	91.37	90.44	91.35	90.98	89.05	90.68	90.48	89.05	87.57
MLP	92.40	89.10	90.09	89.57	88.47	89.07	89.46	89.57	88.66
NBTree	87.09	87.23	88.36	85.48	84.20	95.35	82.93	80.58	81.59

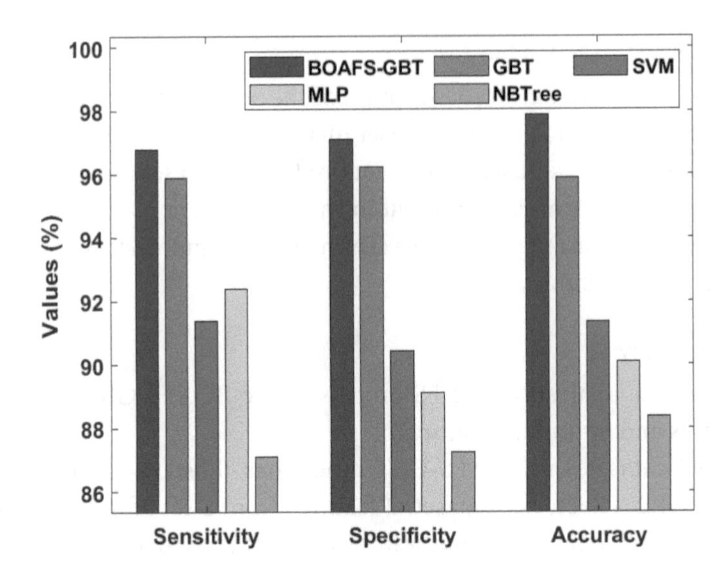

FIGURE 3.3 Classification analysis of BOAFS-GBT model on GPS trajectories dataset.

higher classifier outcome with the sensitivity of 91.37%, specificity of 90.44%, and accuracy of 91.35%. Likewise, the GBT model has shown moderate performance with the sensitivity of 95.88%, specificity of 96.23%, and accuracy of 95.89%. Finally, the BOAFS-GBT model has shown superior results with the maximum sensitivity of 96.78%, specificity of 97.10%, and accuracy of 97.88%.

On the applied movement prediction dataset, the NBTree method has reached considerable results with the lower sensitivity of 85.48%, specificity of 84.20%, and accuracy of 95.35%. Simultaneously, the MLP method has showcased tremendous outcome by achieving better sensitivity of 89.57%, specificity of 88.47%, and accuracy of 89.07%. Similarly, the SVM approach has illustrated manageable classifier outcome with the sensitivity of

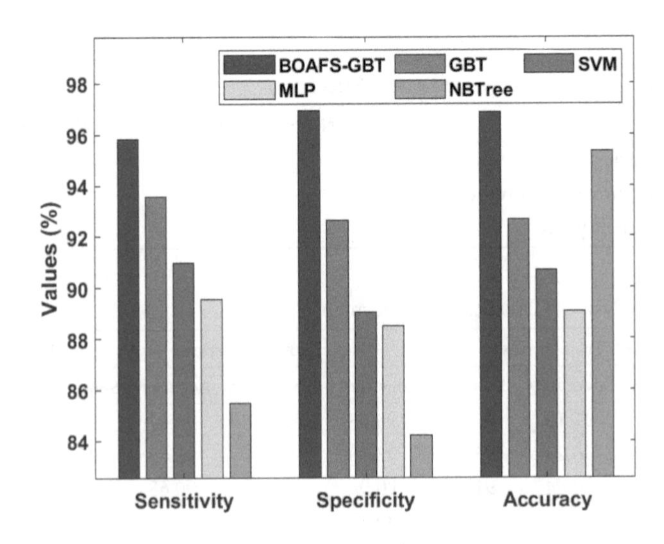

FIGURE 3.4 Classification analysis of BOAFS-GBT model on movement prediction dataset.

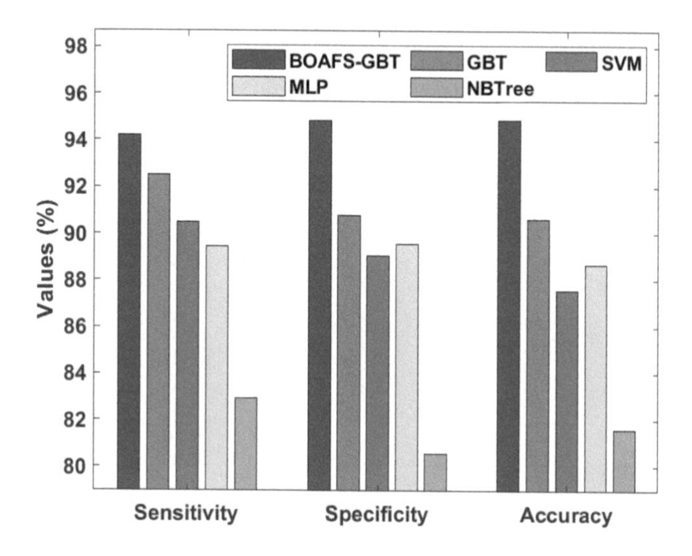

FIGURE 3.5 Classification analysis of BOAFS-GBT model on water treatment dataset.

90.98%, specificity of 89.05%, and accuracy of 90.68%. Along with that, the GBT technology has demonstrated gradual function with the sensitivity of 93.58%, specificity of 92.64%, and accuracy of 92.65%. Finally, the BOAFS-GBT approach has depicted supreme outcome with the higher sensitivity of 95.83%, specificity of 96.92%, and accuracy of 96.87%.

Similarly, on the applied water treatment dataset, the NBTree technique has accomplished better outcomes with the lower sensitivity of 82.93%, specificity of 80.58%, and accuracy of 81.59%. The MLP method has illustrated remarkable outcome by attaining moderate sensitivity of 89.46%, specificity of 89.57%, and accuracy of 88.66%. Similarly, the SVM approach has depicted gradual classifier results with the sensitivity of 90.48%, specificity of 89.05%, and accuracy of 87.57%. Likewise, the GBT approach has demonstrated reasonable function with the sensitivity of 92.49%, specificity of 90.79%, and accuracy of 90.63%. Consequently, the BOAFS-GBT technology has provided best outcomes with the higher sensitivity of 94.20%, specificity of 94.82%, and accuracy of 94.88%.

3.4.3 Energy Consumption Analysis

Table 3.4 and Figure 3.6 provide an energy consumption analysis of the presented model with and without the optimization model. Under the data size of 50, the Hadoop without

TABLE 3.4 Energy Consumption (J) Analysis

Data Size	Hadoop with Optimization	Hadoop without Optimization
50	125	180
100	160	210
150	180	240
200	235	255
250	260	290
300	295	320

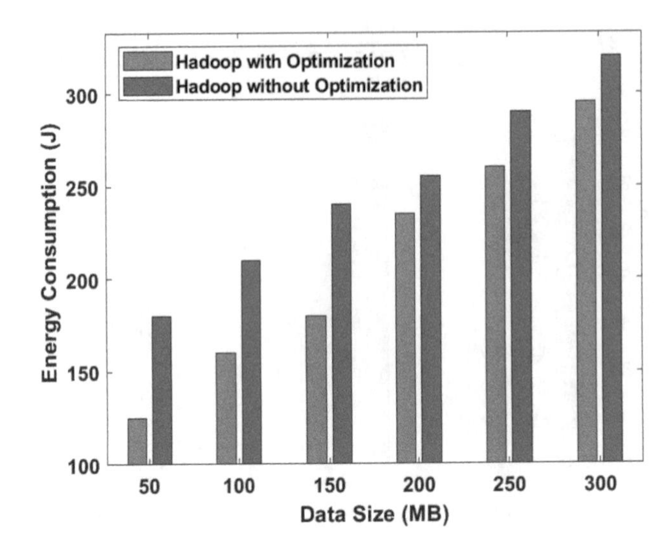

FIGURE 3.6 Energy consumption analysis with and without the optimization model.

optimization requires higher energy consumption of 180 J, whereas the Hadoop with optimization necessitate lower energy consumption of 125 J. Initially, under the data size of 100, the Hadoop with no optimization needs maximum power application of 210 J, whereas the Hadoop with optimization requires least energy application of 160 J.

Next, under the data size of 150, the Hadoop that lacks optimization demands for greater energy consumption of 240 J, whereas the Hadoop with optimization needs minimal energy consumption of 180 J. Then, under the data size of 200, the Hadoop with no optimization necessitate greater power utilization of 255 J, whereas the Hadoop with optimization requires minimum power application of 235 J. Afterward, under the data size of 250, the Hadoop without optimization acquires maximum power utilization of 290 J, whereas the Hadoop with optimization requires low power application of 260 J. Simultaneously, under the data size of 300, the Hadoop with no optimization needs maximum power consumption of 320 J, whereas the Hadoop with optimization requires minimum energy consumption of 295 J.

3.4.4 Throughput Analysis

Table 3.5 and Figure 3.7 provide the throughput analysis of the presented model with and without the optimization model. Under the data size of 50, the Hadoop without

TABLE 3.5 Throughput (Kbps) Analysis

Data Size	Hadoop with Optimization	Hadoop without Optimization
50	92	86
100	95	89
150	96	91
200	94	88
250	98	93
300	97	92

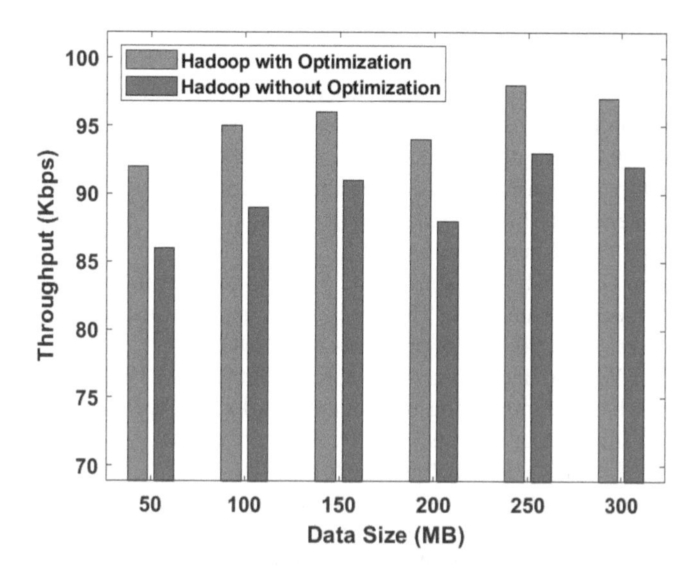

FIGURE 3.7 Throughput analysis with and without the optimization model.

optimization achieves minimal throughput of 86 Kbps, whereas the Hadoop with optimization offers higher throughput of 92 Kbps. Under the data size of 100, the Hadoop that lacks optimization accomplishes lower throughput of 89 Kbps, whereas the Hadoop with optimization provides maximum throughput of 95 Kbps. Next, under the data size of 150, the Hadoop with no optimization attains lower throughput of 91 Kbps, whereas the Hadoop with optimization gives better throughput of 96 Kbps. Similarly, under the data size of 200, the Hadoop with no optimization obtains least throughput of 88 Kbps, whereas the Hadoop with optimization produces greater throughput of 94 Kbps.

Likewise, under the data size of 250, the Hadoop with no optimization reaches lower throughput of 93 Kbps, whereas the Hadoop with optimization gives maximum throughput of 98 Kbps. Finally, under the data size of 300, the Hadoop with no optimization attains lower throughput of 92 Kbps, whereas the Hadoop with optimization generates remarkable throughput of 97 Kbps.

3.5 CONCLUSION

This chapter has introduced an effective big data analytics method in SIoT using BOAFS-GBT model. Once the data is generated by SIoT devices, several transformations occur by utilization of transmit engines such as moving, cleaning, splitting, and merging. Thereafter, the data would be saved in several methods such as cloud or different databases. Afterward, the proposed BOAFS-GBT model primarily performs feature selection using BOAFS model, which selects a constructive collection of features from the big data. Next, the GBT model is used for the classification of the feature-reduced data into several classes. The results of the BOAFS-GBT model has been validated against three datasets and the obtained results notified the effective performance by attaining maximum accuracy of 97.88% on GPS trajectories dataset, 96.87% on movement prediction dataset, and 94.88% on water treatment dataset. In future, the results can be further improvised by the use of clustering techniques.

REFERENCES

1. Xuanxia Yao, Zhi Chen, and Ye Tian, "A lightweight attribute-based encryption scheme for the Internet of Things," *Future Gener. Comput. Syst.*, vol. 49, pp. 104–112, Aug. 2015. http://dx.doi.org/10.1016/j.future.2014.10.010.
2. Zheng Yan, Jun Liu, Athanasios V. Vasilakos, and Laurence T. Yang, "Trustworthy data fusion and mining in Internet of Things," *Future Gener. Comput. Syst.*, vol. 49, pp. 45–46, Aug. 2015. http://dx.doi.org/10.1016/j.future.2015.04.001.
3. L. Atzori, A. Iera, G. Morabito, and M. Nitti, "The Social Internet of Things (SIOT) - when social networks meet the Internet of Things: concept, architecture and network characterization," *Comput. Netw.*, vol. 56, no. 16, pp. 3594–3608, 2012.
4. Ruixin Maa, Kai Wanga, Tie Qiua, Arun Kumar Sangaiahb, Dan Lina, and Hannan Bin Liaqatc, "Feature-based compositing memory networks for aspect-based sentiment classification in Social Internet of Things," *Future Gener. Comput. Syst.*, vol. 92, pp. 879–888, Mar. 2019. http://dx.doi.org/10.1016/j.future.2017.11.036.
5. Muhammad Babar and Fahim Arif, "Smart urban planning using big data analytics to contend with the interoperability in Internet of Things," *Future Gener. Comput. Syst.*, vol. 77, pp. 65–76, December 2017. http://dx.doi.org/10.1016/j.future.2017.07.029.
6. Bo Yuan, Lu Liu, and Nick Antonopoulos, "Efficient service discovery in decentralized online social networks," *Future Gener. Comput. Syst.*, vol. 86, pp. 775–791, September 2017. https://doi.org/10.1016/j.future.2017.04.022.
7. Mohammed Zaki Hasan and Fadi Al-Turjman, "SWARM-based data delivery in social Internet of Things," *Future Gener. Comput. Syst.*, vol. 92, pp. 821–836, Mar. 2019. http://dx.doi.org/10.1016/j.future.2017.10.032.
8. G. Han, L. Zhou, H. Wang, W. Zhang, and S. Chan, "A source location protection protocol based on dynamic routing in WSNs for the social Internet of Things," *Future Gener. Comput. Syst.*, vol. 82, pp. 689–697, May 2018.
9. J. Lopez, R. Rios, F. Bao, and G. Wang, "Evolving privacy: from sensors to the Internet of Things," *Future Gener. Comput. Syst.*, vol. 75, pp. 46–57, Oct. 2017.
10. A. Paul, A. Ahmad, M. M. Rathore, and S. Jabbar, "Smartbuddy: defining human behaviors using big data analytics in Social Internet of Things," *IEEE Wireless Commun.*, vol. 23, no. 5, pp. 68–74, Oct. 2016.
11. S. F. Ochoa, G. Fortino, and G. Di Fatta, "Cyber-physical systems, Internet of Things and big data," *Future Gener. Comput. Syst.*, vol. 75, pp. 82–84, Oct. 2017.
12. M. Wasi-ur-Rahman et al., "High-performance RDMA-based design of Hadoop Map Reduce over InfiniBand," in *Proc. IEEE 27th Int. Parallel Distrib. Process. Symp. Workshop Ph.D. Forum (IPDPSW)*, pp. 1908–1917, May 2013.
13. E. Ahmed et al., "The role of big data analytics in Internet of Things," *Comput. Netw.*, vol. 129, pp. 459–471, Dec. 2017.
14. M. M. Rathore, A. Ahmad, and A. Paul, "IoT-based smart city development using big data analytical approach," in *Proc. IEEE Int. Conf. Autom. (ICA-ACCA)*, Oct. 2016, pp. 1–8.
15. F. Alam, R. Mehmood, I. Katib, and A. Albeshri, "Analysis of eight data mining algorithms for smarter Internet of Things (IoT)," *Proc. Comput. Sci.*, vol. 98, pp. 437–442, Jan. 2016.
16. C.-M. Huang, C.-H. Shao, S.-Z. Xu, and H. Zhou, "The Social Internet of Thing (S-IOT)-based mobile group handoff architecture and schemes for proximity service," *IEEE Trans. Emerg. Topics Comput.*, vol. 5, no. 3, pp. 425–437, Jul./Sep. 2017.
17. M. Z. Hasan and F. Al-Turjman, "SWARM-based data delivery in Social Internet of Things," *Future Gener. Comput. Syst.*, vol. 92, pp. 821–836, Mar. 2019.
18. A. Ahmad et al., "Toward modeling and optimization of features selection in big data based Social Internet of Things," *Future Gener. Comput. Syst.*, vol. 82, pp. 715–726, May 2017.

19. W. Mardini, Y. Khamayseh, M. B. Yassein, and M. H. Khatatbeh, "Mining Internet of Things for intelligent objects using genetic algorithm," *Comput. Electr. Eng.*, vol. 66, pp. 423–434, Feb. 2017.
20. Bao Le Nguyen, E. Laxmi Lydia, Mohamed Elhoseny, Irina V. Pustokhina, Denis A. Pustokhin, Mahmoud Mohamed Selim, Gia Nhu Nguyen, and K. Shankar, "Privacy preserving blockchain technique to achieve secure and reliable sharing of IoT data," *Comput. Mater. Continua*, vol. 65, no. 1, pp. 87–107, July 2020.
21. Mohamed Elhoseny, Mahmoud Mohamed Selim, and K. Shankar, "Optimal deep learning based convolution neural network for digital forensics face sketch synthesis in Internet of Things (IoT)," *Int. J. Mach. Learn. Cybern*, July 2020. https://doi.org/10.1007/s13042-020-01168-6
22. Sachi Nandan Mohanty, K. C. Ramya, S. Sheeba Rani, Deepak Gupta, K. Shankar, S. K. Lakshmanaprabu, and Ashish Khanna, "An efficient lightweight integrated block chain (ELIB) model for IoT security and privacy," *Future Gener. Comput. Syst.*, vol. 102, pp. 1027–1037, Jan. 2020.

CHAPTER 4

An Energy-Efficient Fuzzy Logic-Based Clustering with Data Aggregation Protocol for WSN-Assisted IoT System

Tran Hoang Vu[1] and Gianhu Nguyen[2,3]

[1]*The University of Danang – University of Technology and Education, Da Nang, Vietnam*

[2]*Faculty of Information Technology, Duy Tan University, Da Nang, Vietnam*

[3]*Graduate School, Duy Tan University, Da Nang, Vietnam*

CONTENTS

4.1 INTRODUCTION

The progressive development of IoT enhances the lifestyle and performance of a human being [1]. IoT is one of the innovative patterns of accumulating different objectives that is composed of two semantics. Initially, the base of IoT is a network that is assumed to be the enhancement of the Internet. Second, the end users have exchanged the data, which is referred to as connectivity over objects. Because of the sensing performance and wireless interaction ability, the usage of IoT provides extensive benefits. As it is a pervasive method, it can also apply in vehicle observation, medicare, urban transportation, space searching, and some other challenging applications [2]. The significant member of IoT is named as WSN, and it is treated as common environment for several fields like estimation of diverse ecological variables such as temperature, pressure, humidity, light, and so on.

This network is applied in massive domains like environmental observation, medical sector, electricity grid, surveillance, and persistent patient observation, respectively. Under the reduction of wiring expenses and activating new kinds of evaluation domains, the remote system enhances the wired structures. Thus, remote monitoring has numerous benefits in several other applications. It is one of the three system topologies gathered in WSN. Each hub interfaces specifically with a path in star topologies [3]. In case of collective network, the tree is organized in such a way that all hub interfaces with a hub maximum in a tree to door. A data has been steered from lower hub on a tree to portal. Consequently, it represented that one hub can interface with diverse hubs for providing upgraded unwavering quality in network.

The WSN is composed of different connection units that is additive with wireless network. A link network is changed in the network node, which develops various modules, for instance, battery, analog circuit, sensor interfaces radio, microcontroller, etc. [4]. In last decades, the Internet is replaced for simulating new ideas regarding IoT from interfacing individuals for connecting the objects. Hence, it has resulted in developing the novel pattern into the Internet and provides best applications and business [5, 6]. Figure 4.1 shows the layers in WSN-assisted IoT systems.

In WSN, the base station (BS) should generate the collected value for end users. Under the reduction of transmission load and significant application, the data collection is forwarded [7]. These hubs are applicable in minimum collections, which support data collection named as clusters. For data accumulation, clustering is assumed to be isolation of hubs [8–10]. Thus, clustering is mainly applied for enhancing the network lifetime. Some important measures are taken for determining the implementation of sensor nodes. In order to accomplish better elasticity of well-organized system, possible productivity is carried out [11, 12]. According to the borders enhancement of group, additional portions are allocated to hubs. The cluster

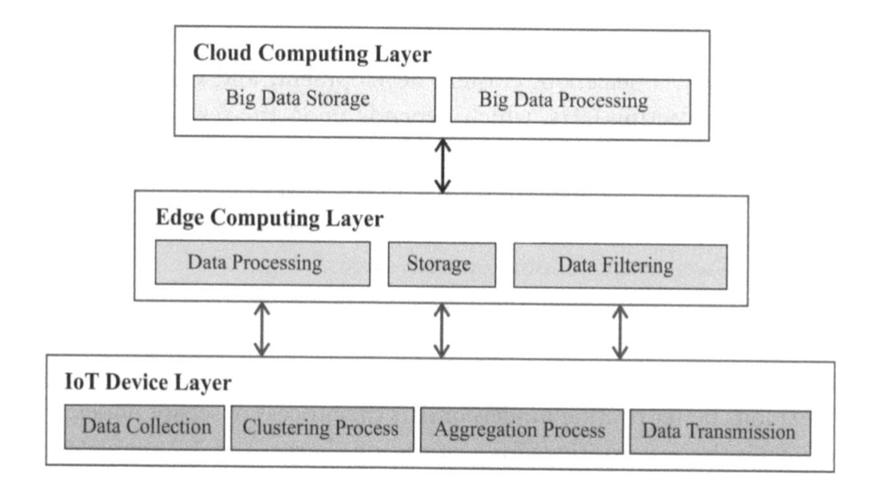

FIGURE 4.1 Layers in WSN-assisted IoT systems.

head (CH) is termed as an incharge of management and sending data to BS present in group. By the replacement of hubs that are applicable for predicting and transmitting, the collected data to CH is termed as cluster member (CM) nodes [13–15].

The sensors are capable of producing massive data and contain various features like processing energy, storage, and communication potentials [16–18]. WSNs are said to be identical when the nodes are symmetric. A WSN that is nonidentical is said to be heterogeneous. Basically, the machines are battery-powered and data collection is highly effective and significant [19–21]. Clustering is a well-known and power-effective solutions developed by the research community for gathering information from the WSN. Consequently, a group of clusters are deployed. A cluster is composed of a member node and CH. It is helpful in gathering data from members called as intracluster communication. CHs support to address the centralized BS named as intercluster communication [23].

Feedback mechanism-based unequal clustering (FMUC) [24] is defined as the feedback approach that depends upon unequal heterogeneous protocol. FMUC is mainly applied for eliminating energy hole problem or hot spot issue, while the energy load is managed in application-based WSNs. First, FMUC separates the system as layers and processed systematically. An arithmetic method has been employed for making equal power application and overall initial energy of all layers. The cluster would be one among these layers. The cluster size can be determined by the amount of power utilization of a layer. Clusters forward the sizes as a feedback to BS where it publishes the gathered values to a network. Every nodes of WSN derives an opinion measure, although only CH modifies the processing requirements on the basis of obtained values.

An unequal HEED (UHEED) [25] is defined as uneven-sized cluster-dependent procedure to WSNs. UHEED applies a strategy of EEUC procedure to HEED for developing unequal size clusters. A cluster size of CH is based on a distance measure. On the other hand, a cluster that is placed distant from BS contains maximum range in terms of clusters closer to sink. UHEED minimizes the hot spot issue and enhances network duration than other models. Rotated UHEED (RUHEED) applies the unequal-sized cluster-dependent model that maximizes the hot spot problem and increases the network survival rate. RUHEED contains three phases: CH selection, clusters development, and CH rotation. HEED is mainly employed for selecting CHs, which depends upon the residual energy (RE) and processing expense. EEUC is relied on the distance from sensor node to BS, which has been utilized for developing unequal-sized clusters. Reclustering of network is carried out if sensor nodes exhaust the complete energy. RUHEED saves power and reduces the count of cluster selection as well as cluster development stages.

In Ref. [26], the diverse energy-yielding models are presented: (i) energy harvesting integrated with concurrent data decoding, which is referred to as a trade-off from energy used for future and the amount of energy that is consumed in signal decoding; (ii) energy-effective task of WSN exploits better routing models or lists the process of sensor nodes; (iii) mobile chargers terminates the best positions for performing charge; and (iv) energy distribution. The objectives (iii) and (iv) are selected in a two-step procedure integrating mobile charger that shifts within the system to discharged CH that alters an energy operating from overcharged CHs and alternate nodes. An energy operating is carried out for all clusters. A CH is selected from nodes with maximum number of neighbors within the inclusion circle. Initially, the mobile charger applies a better path to obtain consecutive discharged CH, and it is terminated and overcharges. The overcharged CHs give energy to cluster members (CM) without any competition, where the count of seller nodes is maximum when compared to count of buyer nodes. Reference [27] applies a cross-layer cooperative TDMA model instead of existing methods for optimizing the CHs transmitting function. A CH function is changed among the nodes with the help of duty cycling for individual energy harvesting abilities. It determines the best number of clusters on the basis of intensity of power source. The protocol depends upon LEACH. The CH option relies on possibility application, which employs duty cycle approach in which a node is not capable to become CH prior to passing duty cycles.

This chapter develops an energy-efficient fuzzy-based clustering and data aggregation protocol called FC-DR for WSN-assisted IoT system. The proposed FC-DR method operates on three major stages: FC-based node clustering, data collection, and error-bounded lossy compression (EBLC) technique-based data aggregation. The EBLC technique aggregates the data before sending it to the cloud server, which is treated as a maximum energy consumption stage among the IoT devices. The efficiency of the presented model has been examined under different aspects. The attained simulation results ensured the goodness of the FC-DR model in terms of several evaluation parameters.

4.2 BACKGROUND INFORMATION

In this section, a brief overview of clustering is described in detail with objectives and characteristics [11].

4.2.1 Clustering Objective

A node is clustered in WSN with several objectives that depend on function necessity [12–14]. The energy maintenance and removing hot spot issue are the most general objectives of clustering. Several of the further objectives are described in the following sections.

4.2.1.1 Scalability

Sensor nodes are used in massive numbers ranging from hundreds to thousands, depending on the function necessity in the current situation. A pattern of routing methods has the capability to work with this massive count of sensor nodes [15]. If node in the cluster requires for transmitting data to a node in one more cluster, the nodes must identify the aspects of the connected obtaining CH. The hierarchical structural design gives scalability in huge scale WSN by separating the sensing domain into several layers, and all layers are again separated into count of clusters. These lead to enhanced scalability and diminish the size of the routing table.

4.2.1.2 Fault-Tolerant

For many purposes, sensors are used in harsh environment (e.g., sensors are dropped from helicopter). These nodes contain enhanced danger of physical damage and failure of nodes. Fault-tolerant nodes are significant in vital function where the loss of any sensor data results in catastrophe [16]. Clustering is the method capable of making a fault-tolerant and protected WSN. A self-organized WSN controls the fault by the procedure of reclustering the network. The reclustering procedure not only enhances the resource burden but also interrupts the present process. Fault tolerance is attained by reclustering, allocating backup CH, depute CH, or rotating CH.

4.2.1.3 Data Aggregation/Fusion

Because the huge counts of sensors sense the similar data in the physical surroundings, there is much possibility of needless data. A data aggregation is an efficient method to keep away useless data broadcast, and it also diminishes a count of broadcasts [17–19]. These methods are a signal procedure technique that aggregates each obtained packet into a resultant packet. These methods amplify the general data and suppress the redundant noise. In WSN, CH carries out data aggregation on data that is obtained from its CMs and forwards the aggregated data to BS using single hop or multihop. So, the count of broadcasts and the entire load of the network are also extensively decreased.

4.2.1.4 Load Balancing

Load balancing plays an essential role in extending the network duration. The load balancing is a critical problem where CHs are chosen from accessible nodes in the network. Uniform load distribution between the CHs is important for avoiding hot spot issue. In a clustering security regular load distribution, each of the CH utilizes around similar count of energy. As an outcome, further energy-capable network is simply obtained [20].

4.2.1.5 Stabilized Network Topology

A node is managed into clusters and CH is chosen from all clusters; CH is dependable to some topology modifies at the cluster level. The CH has the data of its CMs such as node id, location and energy level. Controlling the network topology in hierarchical structural design is optimal than flat design. If a node expires or goes to other cluster, this alters are directly registered and informed by CH to BS and reclustering is completed for maintaining the network topology efficiently.

4.2.1.6 Increased Lifetime

An important function of clustering is to enhance the network duration as a long as feasible. Because the sensors are energy-limited, maximizing the network duration is very essential to concurrent functions. Intracluster communication is diminished by choosing nodes as CHs with further neighbor nodes. A clustering and routing procedure can also be joined to maximize duration. A clustering extends the duration of WSN by rotating CHs suitably between the CMs, sleep modes and cluster conservation methods are accurately employed for enhancing the network duration.

4.2.2 Clustering Characteristics

For classifying several clustering manners, many clustering features are utilized. In this section, three characteristics of clustering are described in detail.

- Cluster properties
- CH properties
- Clustering procedure properties

4.2.2.1 Cluster Properties

The design of the clusters is determined as cluster properties that contain cluster count, cluster size, intracluster communication, and intercluster communication [21, 22].

4.2.2.1.1 Cluster Count The count of cluster created is fixed or variable depends on the function necessity. In few cases, the count of cluster is 5% of entire count of the nodes utilized. In several functions, the count of clusters is variable if the CHs are arbitrarily chosen.

4.2.2.1.2 Cluster Size A cluster size is classified into equal and unequal size cluster. In equal clustering, the size of the cluster is similar throughout the network. In unequal clustering, the cluster size is defined depending on the distance to BS. A cluster size is lesser if the distance to BS is small and the size enhances as the distance to BS enhances.

4.2.2.1.3 Intracluster Communication Intracluster communication contains the data broadcast among CH as well as CMs in a cluster. According to the clustering manners, the communication is direct or multihop communication. In huge-scale WSN, multihop communication is required to data broadcast in a cluster.

4.2.2.1.4 Intercluster Communication Intercluster communication is direct or multihop communication. Generally, multihop method is selected for energy-capable data broadcasts from CHs to BS with in-between CHs in huge-scale WSN. A few functions of tiny-scale WSN, such as the communication between CHs and BS, are single-hop broadcast.

4.2.2.2 CH Properties

The CH carries out the subsequent processes: gathering data from CMs, data aggregating and forwarding to BS with direct or multihop communication. The CH obtains data from its CMs, execution data aggregation of the gathered sensor data transmits the combined data to BS.

4.2.2.2.1 Clustering Process A characteristic of clustering procedure is listed under.

4.2.2.2.2 Clustering Methods There are two techniques of clustering: centralized and distributed. In centralized methods, a central authority such as BS or supernodes manages the whole function (cluster formation, CH selection, and so on.) as distributed manners have no central influence and are extremely used in huge-scale WSN.

4.2.2.2.3 Objective of Node Grouping Many objectives of nodes grouping have already been explained in this study, for example, fault tolerance, load balancing, and so on.

4.2.2.2.4 Nature The clustering procedure is proactive, reactive, or hybrid in nature. A node always senses the data and forwards it to CH. In proactive form, the CH always broadcasts the data to BS. In reactive form, CH broadcasts the data every time the sensed value crosses the existing threshold. In hybrid cases, CH broadcasts the data to BS at longer usual time intervals and also if the value crosses the threshold value.

4.2.2.2.5 CH Selection There are three methods for selecting CH in WSN: probabilistic techniques are attributed to dependent technique and preset form. In probabilistic manner, the CHs are chosen arbitrarily without some preceding consideration. In attribute-based technique, several metrics are utilized for selecting CHs such as remaining energy, node degree, node centrality, usual remaining energy, distance to BS, and so on. In preset form, CHs are predetermined before placing the sensors in the sensing domain.

4.3 PROPOSED FUZZY-BASED CLUSTERING AND DATA AGGREGATION (FC-DR) PROTOCOL

In this section, the working principle of the FC-DR protocol has been presented. The basic architecture of clustering-based IoT-WSN model is depicted in Figure 4.2. The presented FC-DR method operates on three major stages: node clustering, data collection, and data aggregation. Initially, the IoT sensor nodes perform FC process to select CHs and organize clusters. In the second level, the CMs observe the environment and forward the data to the CH. Finally, in the third level, the CHs perform data aggregation using EBLC technique.

4.3.1 Fuzzy-Based Clustering Process

For reduced energy utilization, the cluster formation process plays an important part. It applies *k*-means clustering technique to the cluster formed. The counts of datasets are

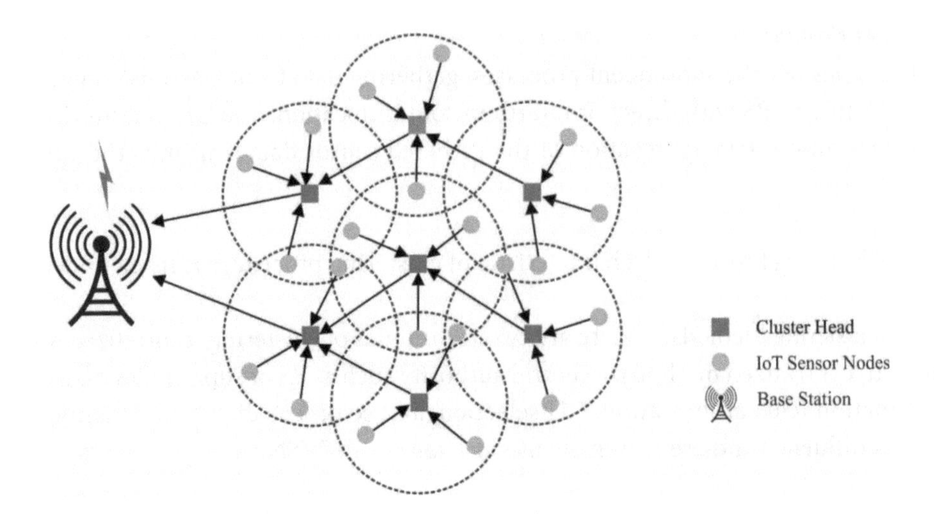

FIGURE 4.2 Architecture of IoT-WSN.

separated into k-clusters that utilize these techniques. A value of k is estimated as given in equation (4.1):

$$k=\sqrt{\frac{n}{2\pi}}\sqrt{\frac{\varepsilon_{fs}}{\varepsilon_{mp}}}\frac{D}{x_{BS^2}} \tag{4.1}$$

where n is the node count of sensor nodes, D is the network size, and x_{BS} is the average distance of every node for the BS. Utilizing the Euclidean distance, the distance among all of the sensors nodes for the entire clusters center is planned as given in equation (4.2):

$$X_{n2CC}=\sqrt{\sum_{j=1}^{N}\left(X_j-X_{CC}\right)} \tag{4.2}$$

where X_{n2CC} indicates node's distance from cluster center, X_j signifies the direct node j, and X_{CC} is the cluster center.

The CH selection takes place using remaining energy, communication rate between a node and its neighboring node, link quality, restart value, and count of neighboring nodes, and node marginality, respectively.

4.3.1.1 Remaining Energy Level

Energy is a significant resource in WSN. The CHs are the nodes that utilize and further energy than CMs if they contain aggregating, computing, and routing information. The remaining energy is calculated as given in equation (4.3):

$$E_r=E_0-E_c \tag{4.3}$$

where E_0 and E_c are the primary energy and the energy utilized with the node, respectively, and E_r is the remaining energy of a standard node.

4.3.1.2 Communication Rate

The broadcasting message utilizes the energy that is simultaneously a square of the distance between the applicant and source nodes. A charge of communication rate is determined as given in equation (4.4):

$$C = \frac{d_{avg}^2}{d_0^2} \tag{4.4}$$

where d_{avg} signifies the average distance among the neighboring nodes and d_0 is the transmitting range of the nodes.

4.3.1.3 Link Quality

In WSN, the disappearing channel is usually arbitrary and time-different. Until a receiver does not examine the signal properly, a rebroadcast can happen and it needs further energy dissipation of the broadcaster. Thus, a link quality should be estimated for accomplishing energy competence. A link quality can be computed as given in equation (4.5):

$$Q = \frac{Q_i - Q_{min}}{Q_{max} - Q_{min}} \tag{4.5}$$

where Q_{max} and Q_{min} are the maximal and minimal count of rebroadcast from the neighborhood, respectively, and Q_i indicates the entire rebroadcast number among the neighbors and the node.

4.3.1.4 Restart Value

A node is essentially an embedding method. Occasionally, the methods affect a software or hardware fault. For solving these issues, the watchdog circuit is employed for restarting the PC scheme to ensure the node continues functioning. So, the frequent restart is utilizing any further energy. A value of restart value is estimated utilizing equation (4.6):

$$S = 1 - \frac{S_0 - S_{min}}{S_{max} - S_{min}} \tag{4.6}$$

where S_{max} and S_{min} are the maximal and minimal restart values obtained from neighboring, nodes, respectively, and S_0 indicates the entire restart value behind the WSN has been arranged.

4.3.1.5 Node Degree

A principle is that the nearer the neighboring nodes, the more effective the node and the greater possibility of becoming a CH. The node degree can be calculated as given in equation (4.7):

$$D = \frac{|D_i - D_0|}{D_0} \tag{4.7}$$

where D_i indicates the count of neighboring nodes and D_0 is the better count of neighboring nodes.

4.3.1.6 Node Marginality

A division of coverage area will incorporate absence of nodes when a node is placed at the boundary of the observing region. A node simply encloses a limited area. Thus, the entire count of CHs in the network gets enhanced. Node marginality is determined as follows:

$$M = \frac{q}{4} \tag{4.8}$$

where q indicates the quadrant number.

In fuzzy surroundings, fuzzy analytic hierarchy process (AHP) is a valuable method under several conditions of decision-making. According to the function objectives and user preferences, the weights are assigned for all conditions in AHP. A fuzzy AHP technique depends upon the following stages:

1. Creation of pairwise comparative decision matrix

   A pairwise comparative matrices are provided as follows:

$$A = \begin{pmatrix} y_{11} & y_{12} & \cdots & \cdots & y_{1n} \\ y_{21} & y_{22} & \cdots & \cdots & y_{2n} \\ \vdots & \vdots & \vdots & \ddots & \vdots \\ y_{n1} & y_{n2} & \cdots & \cdots & y_{nn} \end{pmatrix} \tag{4.9}$$

   where $y_{ii} = 1$, $y_{ji} = 1 / y_{ji}$.

2. Normalization of decision matrix as computed in equation (4.10)

$$a_{ij} = \frac{y_{ij}}{\sum_{i=1}^{n} y_{ij}} \tag{4.10}$$

3. Weighted normalized decision matrix as provided in equation (4.11)

$$W_i = \frac{\sum_{i=1}^{n} a_{ij}}{n} \text{ and } \sum_{i=1}^{n} W_i = 1 \tag{4.11}$$

where n is the criterion number.

4.3.2 Data Aggregation Process

The EBLC method such as SZ performs high-performance computing (HPC) functions. These compression methods have been presented to manage the massive counts of data created in the implementation of HPC functions. The actual SZ reduces input data records, which are in binary formats and contain several data shapes as well as types. It is presented for adapting the SZ technique to IoT tools by considering the floating point data type and removing other types that create the code tiny in size and are simple for compiling on small tools. Besides, the technique was modified for taking a 1D arrangement of float sensor data as input and replacing a byte range, which exists to be broadcasted for the edge node. After selecting, SZ for IoT functions as follows:

- SZ access the encoding of multivariate time series, including different features through several scales.
- SZ permits managing the data loss by utilizing an error bound.
- SZ takes superior compression ratio compared to the multidimensional change field compression methods

It is regarded that the information are broadcasted for the edge behind all periods P of time t. The gathered information is in the structure of $M \times N$ array, where M indicates the count of readings and N refers to the count of features. In the beginning, the 2D array is changed to the 1D array. Thereafter, the flattened array compresses utilizing the lossy SZ method. Finally, the outcome binary array is broadcasted for the edge. It is noticeable that the adaption has been completed through removing the essential performances from the actual SZ for making it suitable on wearable and resource-limited tools.

An SZ compressing technique begins with reducing the 1D array utilizing adaptive curve-fitting methods. The optimal fit stage uses three forecast methods: preceding neighbor fitting (PNF), linear curve-fitting (LCF), and quadratic curve-fitting (QCF). A variation among the three methods is found in the count of precursor data points needed to suit the actual rate. An adapted method is the one that gives the nearer estimate. It is to be noticed that the suitable information is changed into integer quantization factors and encoding that utilize Huffman tree. If no forecast methods in the curve-fitting stage ensure the error limit, the data point is clear as uncertain and the next encoding examines the IEEE 754 binary illustration. In the error bound, an absolute error bound (AEB) is utilized, implying that the compressing or decompressing faults are restricted to be in an AEB. For example, when the rate of data point is regarded to be X, an AEB of 10^{-1} implies, and thus the decompressed rate might be in the series $[X-10^{-1}, X+10^{-1}]$.

4.4 PERFORMANCE VALIDATION

In this section, a comprehensive experimental validation of the FC-DR model takes place under diverse ways. Figure 4.3 depicts the energy consumption analysis of the FC-DR method. The figure shows that the FRLDG model has offered maximum energy utilization, but has appeared as an ineffective model. Simultaneously, the MOBFO-EER model has attained slightly higher energy consumption over FRLDG model, but it is not lower than FEEC-IIR model and FC-DR models. On the other hand, the FEEC-IIR model has attained even lower energy consumption over the compared methods. But the presented FC-DR model has attained least energy consumption and appeared as an effective model under varying number of IoT nodes.

Figure 4.4 demonstrates the network existence analysis of the FC-DR method. The figure portrayed that the FRLDG method has provided lower network duration and is represented as worst method. At the same time, the MOBFO-EER approach has accomplished medium network existence than FRLDG method; however, it is not greater than FEEC-IIR and FC-DR methodologies. Besides, the FEEC-IIR scheme has reached slightly better network duration when compared with earlier technologies. However, the projected FC-DR approach has accomplished remarkable network duration and is considered as optimal model under diverse count of IoT nodes.

Figure 4.5 illustrates the PDR analysis of FC-DR approach. The figure represented that the FRLDG framework has provided least PDR and is depicted as worst technique. The MOBFO-EER scheme has reached minimal PDR over FRLDG model; it is not greater than FEEC-IIR and FC-DR technologies. Then, the FEEC-IIR framework has accomplished considerable

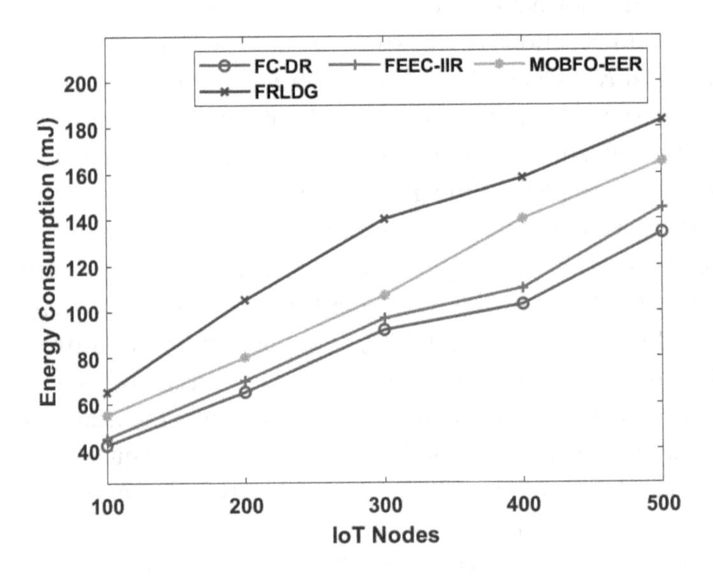

FIGURE 4.3 Energy consumption analysis of the FC-DR model.

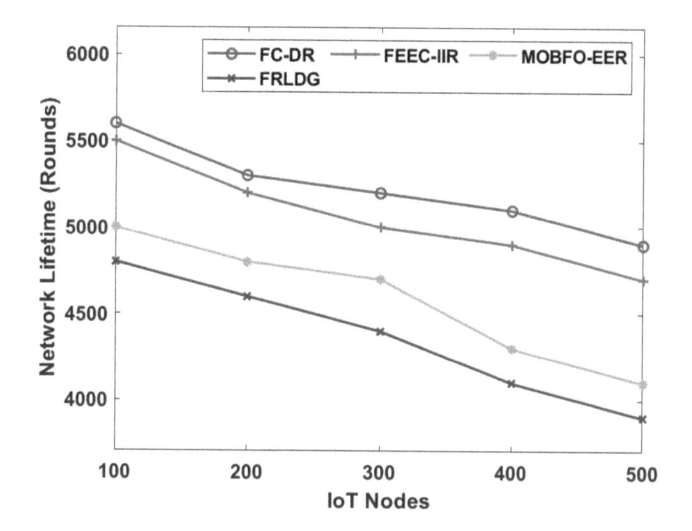

FIGURE 4.4 Network lifetime analysis of the FC-DR model.

PDR than earlier approaches. Therefore, the proposed FC-DR scheme has achieved higher PDR and demonstrated as effectual technology even under different IoT nodes.

Figure 4.6 illustrates the throughput analysis of the FC-DR technique. The figure implied that the FRLDG method has provided higher throughput and is considered as an ineffective approach. Concurrently, the MOBFO-EER framework has achieved minimum throughput than FRLDG model, but it does not exceed FEEC-IIR and FC-DR approaches. The FEEC-IIR technique has obtained slightly better throughput over the earlier methods. Hence, the projected FC-DR scheme has reached a higher throughput and showcased optimal performance under various numbers of IoT nodes.

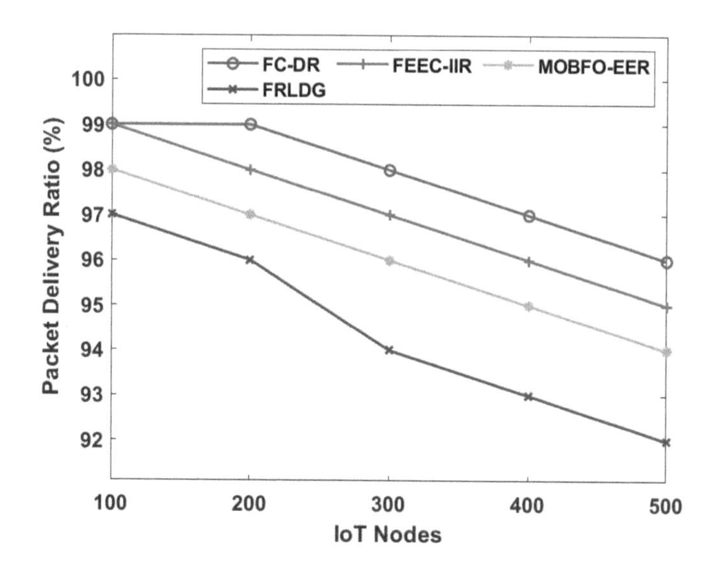

FIGURE 4.5 PDR analysis of the FC-DR model.

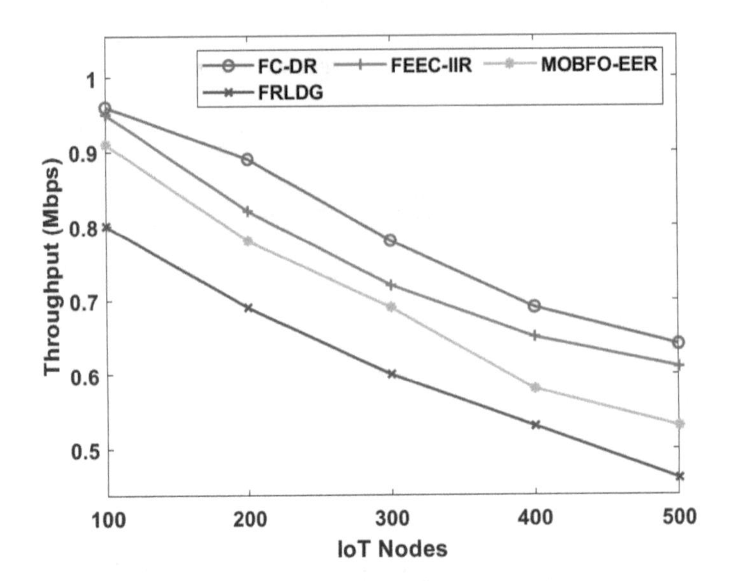

FIGURE 4.6 Throughput analysis of the FC-DR model.

Figure 4.7 represents the end-to-end delay analysis of the FC-DR scheme. The figure portrayed that the FRLDG approach has provided greater end-to-end delay and is considered as a worst technology. The MOBFO-EER scheme has reached better end-to-end delay over FRLDG framework, which does not exceed FEEC-IIR and FC-DR frameworks. The FEEC-IIR method has minimum end-to-end delay than the traditional models. Thus, the projected FC-DR approach has reached lower end-to-end delay and is shown as a best model under distinct number of IoT nodes.

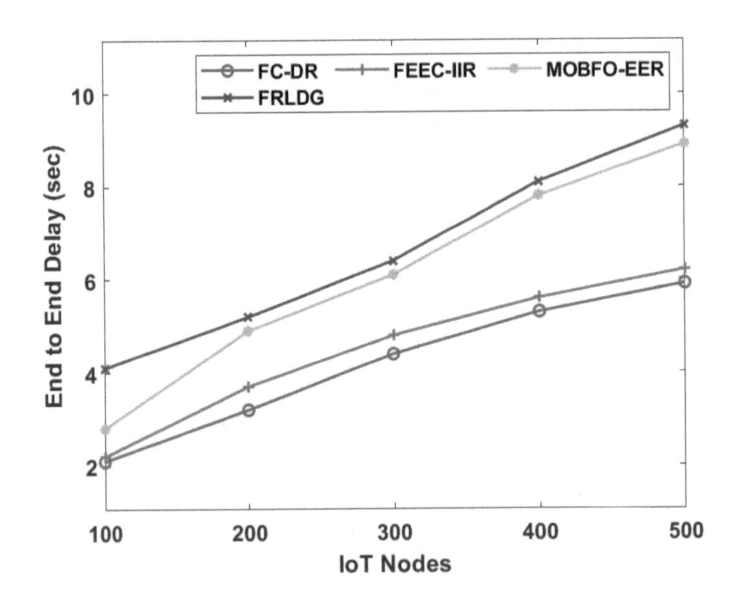

FIGURE 4.7 End-to-end delay analysis of the FC-DR model.

4.5 CONCLUSION

This chapter has developed an energy-efficient FC-DR model for WSN-assisted IoT system. The proposed FC-DR method operates on three major stages: FC-based node clustering, data collection, and EBLC technique-based data aggregation. Initially, the IoT sensor nodes perform FC process to select CHs and organize clusters. In the second level, the CMs observe the environment and forward the data to the CH. Finally, in the third level, the CHs perform data aggregation using EBLC technique. The experimental results denoted that the FC-DR model has shown superior performance by attaining maximum energy efficiency and network lifetime. In future, the performance of the FC-DR can be further improved using routing techniques.

REFERENCES

1. Aierken, N., Gagliardi, R., Mostarda, L., Ullah, Z. (2015). Ruheed-Rotated Unequal Clustering Algorithm for Wireless Sensor Networks. In *29th IEEE International Conference on Advanced Information Networking and Applications Workshops, AINA 2015 Workshops*, March 24–27, 2015, (pp. 170–174), Gwangju, South Korea.
2. Alsheikh, M. A., Lin, S., Niyato, D., Tan, H. P. (2014). Machine learning in wireless sensor networks: algorithms, strategies, and applications. *IEEE Commun Surv Tutor* 16(4):1996–2018, Fourthquarter.
3. Khan, F. A., Ahmad, A., Imran, M. (2018). Energy optimization of PR-LEACH routing scheme using distance awareness in Internet of Things networks. *Int J Parallel Progr* 56:1–20.
4. Kuo, Y.-W., Li, C.-L., Jhang, J-H., Lin, S. (2018). Design of a wireless sensor network-based IoT platform for wide area and heterogeneous applications. *IEEE Sens J* 18(12):5187–5197.
5. Farman, H., Jan, B., Javed, H., Ahmad, N., Iqbal, J., Arshad, M., Ali, S. (2018). Multi-criteria based zone head selection in Internet of Things based wireless sensor networks. *Future Gener Comput Syst* 87:364–371.
6. Mahajan, S., Dhiman, P. K. (2016). Clustering in WSN: a review. *Int J Adv Res Comput Sci* 7(3):198–201.
7. Ramachandran, N., Perumal, V. (2018). Delay-aware heterogeneous cluster-based data acquisition in Internet of Things. *Comput Electr Eng* 65:44–58.
8. Aadri, A., Idrissi, N. (2017). An Energy Efficient Hierarchical Routing Scheme for Wireless Sensor Networks. In *Computer Science & Information Technology* (pp. 137–148).
9. Singh, M., Kumar, N. (2018). Energy efficient routing protocol in IoT. *J Netw Commun Emerg Technol* 8:5.
10. Rajpoot, P., Dwivedi, P. (2019). Multiple parameter based energy balanced and optimized clustering for WSN to enhance the lifetime using MADM approaches. *Wirel Pers Commun* 106(2):829–877.
11. Arjunan, S., Pothula, S. (2019). A survey on unequal clustering protocols in wireless sensor networks. *J King Saud Univ Comput Inform Sci* 31(3), 304–317.
12. Arjunan, S., Sujatha, P. (2018). Lifetime maximization of wireless sensor network using fuzzy based unequal clustering and ACO based routing hybrid protocol. *Appl Intell* 48(8):2229–2246.
13. Arjunan, S., Pothula, S., Ponnurangam, D. (2018). F5N-based unequal clustering protocol (F5NUCP) for wireless sensor networks. *Int J Commun Syst* 31(17):e3811.
14. Uthayakumar, J., Vengattaraman, T., Dhavachelvan, P. (2019). A new lossless neighborhood indexing sequence (NIS) algorithm for data compression in wireless sensor networks. *Ad Hoc Netw* 83:149–157.
15. Uthayakumar, J., Vengattaraman, T., Amudhavel, J. (2017). A simple data compression algorithm for anomaly detection in wireless sensor networks. *Int J Pure Appl Math* 117(19):403–410.

16. Uthayakumar, J., Vengattaraman, T., Amudhavel, J. (2017). A simple lossless compression algorithm in wireless sensor networks: an application of wind plant data. *IIOAB J* 8(2):281–288.
17. Shanmukhi, M., Patil, A., Asif, S., Amudhavel, J. (2018). A novel hybrid cluster based protocols for wireless sensor networks. *Int J Pure Appl Math* 119(14):479–488.
18. Shanthi, G., Sundarambal, M. (2019). FSO-PSO based multihop clustering in WSN for efficient medical building management system. *Cluster Comput* 22(5):12157–12168.
19. Parwekar, P. (2020). SGO A New Approach for Energy Efficient Clustering in WSN. In *Sensor Technology: Concepts, Methodologies, Tools, and Applications* (pp. 716–734), IGI Global.
20. Saihood, A.A., Hasan, Z.S. (2019). Enhanced WOA for mobile energy efficient and delay aware clustering in WSN. *Int J Adv Res Comput Sci* 10(5):8.
21. Mehra, P.S., Doja, M.N., Alam, B. (2020). Fuzzy based enhanced cluster head selection (FBECS) for WSN. *J King Saud Univ Sci* 32(1):390–401.
22. Kumaratharan, N., Padmapriya, N., Dharani, A. (March 2019). A Survey on Improved PSO Routing and Clustering in WSN. In *2019 IEEE International Conference on System, Computation, Automation and Networking (ICSCAN)* (pp. 1–5), IEEE.
23. Sharma, R., Vashisht, V., Singh, U. (2019). EEFCM-DE: energy-efficient clustering based on fuzzy C means and differential evolution algorithm in WSNs. *IET Communications* 13(8):996–1007.
24. Liu T., Peng J., Yang J., Chen G., Xu W. (2017). Avoidance of energy hole problem based on feedback mechanism for heterogeneous sensor networks. *Int J Distrib Sensor Netw* 13(6), doi: https://doi.org/10.1177/1550147717713625.
25. Ever, E., Luchmun, R., Mostarda, L., Navarra, A., Shah, P. (2012). UHEED: An Unequal Clustering Algorithm for Wireless Sensor Networks. In *Proceedings of the 1st International Conference on Sensor Networks – Volume 1: SENSORNETS* (pp. 185–193).
26. Bahbahani, M. S., Alsusa, E. (2018). A cooperative clustering protocol with duty cycling for energy harvesting enabled wireless sensor networks. *IEEE Trans Wireless Commun* 17(1):101–111.
27. Moraes, C., Har, D. (2017). Charging distributed sensor nodes exploiting clustering and energy trading. *IEEE Sensors J* 17(2):546–555.

CHAPTER 5

Analysis of Smart Home Recommendation System from Natural Language Processing Services with Clustering Technique

N. Krishnaraj,[1] R. Lenin Babu,[2] R. Adaline Suji,[3] and Andino Maseleno[4]

[1]Department of Computer Science and Engineering, SRM Institute of Science and Technology, Kattankulathur, Tamil Nadu, India

[2]Conversight.ai, Indiana, Carmel, IN, USA

[3]Department of Computer Science and Engineering, Kalaignar Karunanidhi Institute of Technology, Coimbatore, Tamil Nadu, India

[4]STMIK Pringsewu, Lampung, Indonesia

CONTENTS

5.1 INTRODUCTION

In the recent past, online services have become progressively useful thus extraordinarily improving the efficiency and quality of programming advancement [1]. A natural language processing service (NLPS), as an obliged part of natural language, controls etymological factors, features key issues, and gives a system to design arrangements; the NLPS maybe could be applied to the yields of contention mining in a general structure of refinement. The key adjustment consolidates the articulation "it is regular that," which is a conceivably "natural" linguistic articulation of defeasibility [2, 3]. However, a main point to worry about is the production of apparatuses for savvy home control and the board that address the necessities of nonspecialized people, so as to make this technology generally accessible to anyone [4]. As a home use of Internet of Things (IoT) paradigm, the smart home commits to developing personal satisfaction for occupants, for example, by observing energy utilization of appliance [5].

In the proposed framework, some essential strategies of NLP, such as tokenization, the expulsion of stop words, and parsing, are utilized to comprehend the voice orders. There has been a huge improvement in the area of mechanization utilizing various conventions for regular reason. It is not unexpected to see that cell phones have become an integral part of majority of the individuals' lives nowadays; henceforth, most of the everyday family unit errands are now carried on with the assistance of cell phone [6]. In a savvy city, it is imperative to address the point of individuals and networks as a major aspect of shrewd urban communities, where the individuals from brilliant urban communities are people, networks, and gatherings [7]. One of the significant functionalities of keen urban areas is the correspondence, which gives the capacity of semantic data trade between every single included gathering [8]. In the beginning periods of NLPS, machine learning algorithms and preset transcribed principles were utilized to create the "translation." Now these are replaced by further developed procedures similar to classifier and clustering methods [9].

In this sense, the IoT offers the capability of endless opportunities for new applications and services in the home setting that empower clients to access and control their home conditions from local and remote areas, so as to complete day-by-day life exercises easily from anywhere [10]. All that recently referenced improves the personal satisfaction of the client while simultaneously empowering energy effectiveness for NLPS model in a smart city that implies brilliant home communications [11]. IoT proactive conduct, setting mindfulness, and collaborative communication abilities of savvy home communication are valuable for analysis. Keen home appliances and commands from application stage yet additionally transmit information to the application stage, in this manner getting to be in a generator and receiver of data.

5.2 REVIEW OF LITERATURES

In 2017, Rani et al. [3] proposed that the client sends a direction through discourse to the mobile device, which translates the message and forwards the fitting order to the particular appliance. In this they proposed the plan on executing four essential home appliances as a "Proof-of-Concept" for this venture, which incorporates fan, light, coffee machine, and door alarms. The voice direction given by the client is deciphered by the mobile device utilizing natural language handling. The mobile device goes about as a focal reassure; it figures out what activity must be carried out by which machine to satisfy the client's solicitation.

To utilize such an arrangement to succeed neural architecture, opening name and space esteem forecast assignments by Mishakova et al. prepared natural language understanding models that don't need adjusted information and can together become familiar with the expectation [12]. The planning phases of the NLPS, which do not require modified details, have been prepared and may, together, become familiar with the schedule, the space mark and the opening name of the projected assignments and the reason for the investigation, as well as the readily accessible data collection. The experiments show that a solitary model that learns on unaligned information is focused with best-in-class models, which rely upon aligned information.

In 2019 Alexakis et al. [12] proposed that the IoT agent coordinates a visit bot that can get content or voice directions utilizing NLP. With the utilization of NLP, home gadgets are more easy to use and controlling them is simpler, since in any event, when an order or question/command is unique in relation to the presets, the framework comprehends the client's desires and reacts in like manner. The most important development is that it incorporates a few outsider application programming interfaces and open-source technologies into one blend, highlighting how another IoT application can be manufactured today utilizing a multitier architecture.

In one study by Noguera-Arnaldos et al. [14], the control and the executives of this assortment of gadgets and interfaces represent to another test for non-master clients, rather than creating their life simpler. A natural language interface for the IoT exploits Semantic Web advances to permit non-master clients to control their home environment through a texting application in a simple and instinctive way.

Jivani et al. [15] proposed centralized administration that enables client to control residential machines and services with voice and furthermore settle on electronic choices for the end client's benefit, for example, observing, developing solace, accommodation, controlling encompassing conditions, and conveying required data at whatever point required. The essential goal is to develop a plenarily helpful voice-based framework that uses artificial intelligence and NLPS to control every household application and services, and furthermore get familiar with the client inclinations after some time, utilizing machine learning algorithms.

Qin, and Guo (2019) proposed [16] the challenges that shrewd city is standing up to as far as semantic archive trade is concerned, and proposed a novel machine natural language mediation (MNLM) structure, which gives a sentence-based machine normal language

(MNL) as a sort of intercession language, where every sentence as a compound idea is a lot of nuclear ideas to be perfect with all dialects accomplishing a worldwide semantic change. The MNL empowers the sentence PC intelligible and reasonable through novel codes, without semantic vagueness, and while MNLM will wipe out the semantic irregularity and develop the accurate importance understanding crosswise over conversational contexts.

5.3 SMART HOME: CLOUD BACKEND SERVICES

A smart home coordinates different electrical appliances in the home and computerizes them with zero or least client mediation. The smart home keeps diverse condition factors present and aides the appliances to work as per the requirements of the client. Considering that the main servers keep running behind the cloud services of the outsiders referenced over, this makes our framework exceptionally adaptable, with no exertion from us, so it can support numerous simultaneous clients. Aside from this, the user interface of our application makes the connection with the framework well-disposed and simple. In general, implementation abuses the run-of-the-mill IoT infrastructures by giving an upgraded client experience and convenience of the fundamental IoT infrastructures with the incorporation of NLP, voice acknowledgment, and numerous different technologies. For NLP investigation in shrewd home applications, the cloud offers a full help of improvement and activities and consistent combination standards, so the total lifecycle of application advancement, construct, test, computerized organization, and the board can be taken care of off-premise in the cloud.

5.3.1 Advantages of This System

These systems will, in general, be magnificently flexible with regards to the settlement of new gadgets and machines and other innovations Along with expanded energy efficiency. Depending on the utilization of keen home innovation, it's conceivable to make space more energy-proficient. This system gives security to clients, controlling and confining access. There is a great deal of control and information security, which can be collected to guarantee maximum security for the put-away information. Cloud computing offers one more bit of leeway of working from anyplace over the globe, as long as one has a web connection.

5.3.2 Internet of Things (IoT)

The IoT is a combination layer that permits to associate a few elements together. These can be physical gadgets or sensors, which produce intriguing information or actuators to control encompassing conditions of the world. In request to build work efficiency and solace of an individual, mechanizing environment is basic. There has been a critical advancement in the field of robotization utilizing various conventions for basic reason. The fundamental model for IoT is shown in Figure 5.1. The IoT is quickly making strides in administration and utilization of most recent remote and wire correspondences. To proficiently use the assets in a smart home condition, all the home gadgets ought to be interconnected and give network to the end client so as to control it from anyplace anytime. IoT is changing over the brilliant urban communities and smart homes from publicity concept into the reality.

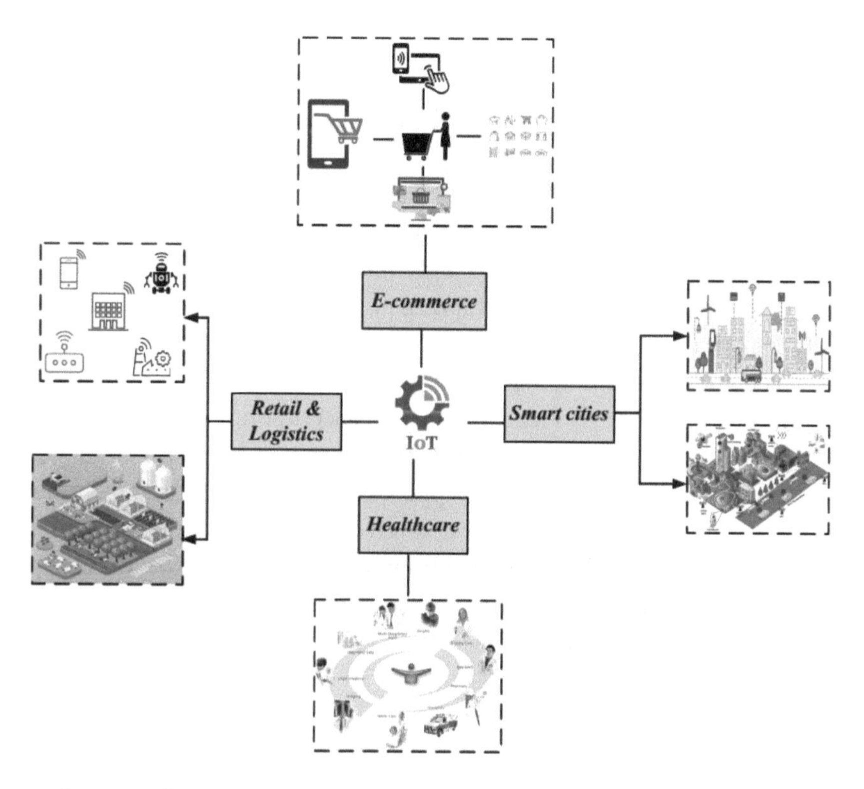

FIGURE 5.1 Structure for IoT.

Smart home implants figuring capacities, networking, and media transmission interfaces in the home apparatuses so as to encourage regular daily existence. The smart home condition includes sensors, actuators, interfaces, and apparatuses arranged together to give restricted and remote control of the environment. Every smart home comprises brilliant items considered sensors and actuators that speak with a focal application. The focal application has different terms in different spaces, for example, a smart home passage. The client's personality ought to be anonymized and access to client's home apparatuses ought to be secured. It is given through secure login and assent of the client. There are shrewd objects and related communication advances to empower an IP-based IoT and the vision of IoT applications.

5.4 OUR PROPOSED APPROACH

A traditional path is to play out a watchword put together inquiry with respect to the portrayals of web-based administrations, which come up short on the capacity to represent complex semantic. The primary propose of the shrewd technology is to give services, for example, health monitoring and mind service, choice passage and specialist co-ops are significant prerequisites to analyze the NLP in savvy home correspondence models. This proposed model comprises various modules such as the recorded voice that will be changed over into content, the word modeling, and stop word evacuation by suggestion inserting modeling. To upgrade the accuracy of proposed model, advanced cluster-generating technique (ACGT) is used. A significant basic subject of this chapter is the interdisciplinary reconciliation of the examination of natural language expressions, a consistent portrayal of the articulations, and a calculable standard language for the coherent representation.

5.4.1 Natural Language Processing Services

NLPS is a documented model of computer science and it serves the client to attempt to state alongside the voice commands. This model, NLP is a field of computer science that encourages us to construe what the client is attempting to state through his voice directions. The NLP in this task gives the client the opportunity to interface with the home machines with his/her own voice and typical language as opposed to muddled PC commands [17]. The greater part of the most recent NLP calculations is based upon machine learning, particularly measurable machine learning. Fundamental purpose of Natural Language Classifier permits identifying the objective from an approaching instant message. Numerous sentences or words which fit the objective must be given so as to make expectation. A noteworthy hidden topic of the work is the interdisciplinary joining of the analysis of normal language expressions, a sensitive portrayal of the expressions, and a calculable standard language for the consistent representation.

5.4.2 Pipeline Structure for NLPS

Preprocessing the sentences in service portrayals utilizing a mediator, here, play out a developed NLP pipeline. This progression needs to mark the limit of a sentence in writings and break them into an accumulation of semantic sentences, which by and large represent intelligent units of thought and will, in general, incorporate an anticipated linguistic structure for further investigation. The outline of NLPS is shown in Figure 5.2. First, named substances are distinguished from writings parsed through NLP pipeline, and afterward the relations that exist between them are extricated.

The client forwards a voice command to the cell device, which translates the message and forwards a suitable direction to the particular machine. The voice order given by the client is deciphered by the cell device utilizing natural language handling. The control situations ought to be planned and written in reasonable design, with completely significant expressions. This is to overcome any issues between natural human reasoning and the

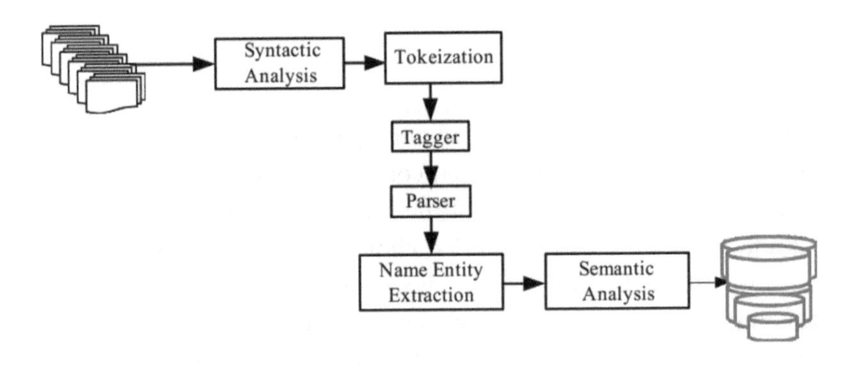

FIGURE 5.2 Pipeline structure for NPLS.

linguistic structure and articulation of the language utilized. It should offer impressive gains in expressiveness and simplicity.

5.4.2.1 Stop Word Removal by Recommendation Systems

The process of converting information to something that a computer can comprehend is alluded to as prepreparing. One of the significant types of prepreparing is to sift through pointless information. In natural language handling, futile words (information) are alluded to as stop words. NLPS are, as a rule, in an unstructured structure. The proposed robotized multireport summarizer is intended to preprocess the crude records; to build up an outline. Under preprocessing, the HTML/XML tags and images are expelled initially. At that point, alongside that, the additional blank area, figures, conditions, and the uncommon characters like "{[(<!@#$%^&*~':+;>)]}?" are likewise evacuated. At long last, the sentence tokenization, stop word removal, and the stemming procedures are likewise executed.

For a lot of records, $H = \{h_1, h_2, \ldots h_n\}$, where "n" means each quantity of records. After the HTML/XML tags and images are evacuated, the sentence segmentation is performed, where every sentence is sectioned separately, $\hat{A} = \left\lfloor A_{h_y 1}, A_{h_y 2,L}, A_{h_y M} \right\rfloor$ from the records. Here, $A_{h_y x}$ denotes x^{th} sentence from the $(d_y)^{th}$ report. Once, the sentences are fragmented, every sentence is tokenized in order to locate the particular words $W = [w_1, w_2, K, w_z]$. Here, 'z' represents each number of unique words.

In addition, from the unique words, the stop words, for example, "an," "a," "the" and so on are expelled, since they have less data about the substance. At last, the stemming procedure is done dependent on Porter Stemming Algorithm, where, the parts of the bargains are cut to change the words to a typical depend structure. As a conclusion of prehandling the online reports, a lot of words, $\tilde{W} = [\tilde{w}_1, \tilde{w}_2, K, \tilde{w}_z]$ are getting for each sentence of the records.

5.4.2.1.1 Frequency of Relevant Term

Frequency of relevant term is the most important highlight utilized so as to rank the sentences. Frequency of relevant term is found for each prehandled sentence that can be assessed by below-given equation (5.1).

$$RT_{freq}\left(\hat{A}_{d_y x}\right) = \max\left(\frac{1}{F_G - 1}\sum_{q=1}^{M_N}\sum_{k=1}^{z}\left(\hat{A}_{d_y x}(p) - A_{d_y x,y}(p)\right)^2\right) \tag{5.1}$$

Here, F_G represents the complete number of sentences from 'N' number of records; 'z' means the number of unique words [18]. In addition, $\hat{A}_{d_y x}(p)$ represents the pth unique term in x^{th} preprocessed sentence from the $(d_y)^{th}$ record and $S_{d_y q}(p)$ is the pth unique term in the other remaining sentences. The term frequency of a word $(\tilde{w}_b)$ is characterized by the number of repetitions that term '$\tilde{w}_b$' attains in the entire set of document; which can be given as in equation (5.2). The second one as inverse sentence frequency, highlight, and the opposite sentence recurrence is a proportion of how much data the world gives, that is, regardless of whether the term is typical or phenomenal in general sentences. The reverse sentence repeat feature can be formed in equation (5.3).

$$\mathrm{TF}(\mathrm{w}_1) = \frac{\mathrm{n}_{(\tilde{\mathrm{w}}_\mathrm{b})}\left\{\hat{\mathrm{A}}_{\mathrm{d}_\mathrm{y}\mathrm{x}}\right\}}{\mathrm{E}} \tag{5.2}$$

$$\mathrm{AF}^{-1}(\tilde{\mathrm{w}}_\mathrm{b}) = \log\left(\frac{\mathrm{E}}{1+\mathrm{E}(\tilde{\mathrm{w}}_\mathrm{b})}\right) \tag{5.3}$$

In equations (5.1) and (5.2) "E" is aggregate of occurrences of each word in every number of unique words and $\hat{A}_{d_y x}$ indicates the prehandled sentence set. The NLPS demonstrating process "Similarity measure" can be utilized to find the proper possibility for the outline by choosing the sentence having the most outrageous similitude with every single other sentence in the information sentence sets. In this way, the Aggregate Cross Sentence Similarity of a sentence $\hat{A}_{d_y x}$ can be computed as,

$$ACS\left(\hat{S}_{d_y x}\right) = \sum_{q-1}^{M_N} Sim\left(\hat{S}_{d_y x}, \hat{S}_{d_y q}\right) \tag{5.4}$$

Where $\hat{S}_{d_y q}$ represents the qth sentence; also $\left(\hat{A}_{d_y x}, \hat{S}_{d_y q}\right) \in \hat{A}_{d_N}$. The above highlights are removed for every one of the sentences in the record set. The positioning of sentences is done depending on the scope of feature vectors so as to make the synopsis. However "stop words" as a rule allude to the most well-known words in a language, there is no single all-inclusive rundown of stop words utilized by all-natural language handling tools, and in fact not all devices even utilize such a list.

5.4.2.2 Word Modeling Procedure

This model offers similar information and yield of a word query table, which originates from word2vect, enabling it to effortlessly supplant then in any network. The info word is decayed into an arrangement of characters $\mathrm{cl}, \mathrm{c2} \ldots \mathrm{c}_n$, where "n" is the length of word. Each character is defined as a one-hot vector 1ci, with one on the index of "c_i" in vocabulary list. This obviously is only a character lookup table, and is utilized to capture similarities between characters in a language. This capacity is mathematically characterized as:

$$\mathrm{L}_\mathrm{i} = \sigma\left(\mathrm{W}_{\mathrm{ri}} * \mathrm{r}_\mathrm{t} + \mathrm{w}_{\mathrm{vi}}\mathrm{v}_{\mathrm{t}-1} + \mathrm{w}_{\mathrm{ci}}\mathrm{c}_{\mathrm{t}-1} + \mathrm{bias}_\mathrm{i}\right) \tag{5.5}$$

The proposed models utilized to register made portrayals out of sentences from words. Be that as it may, the connection between the implications of individual words and the composite meaning of an expression or sentence is seemingly more customary than the relationship of portrayals of characters and the significance of a word. Language demonstrating is an errand with numerous applications in NLPS, by and large, proposed engineering appears in Figure 5.3. An effective language demonstrating requires syntactic parts of language to be displayed, for example, word orderings.

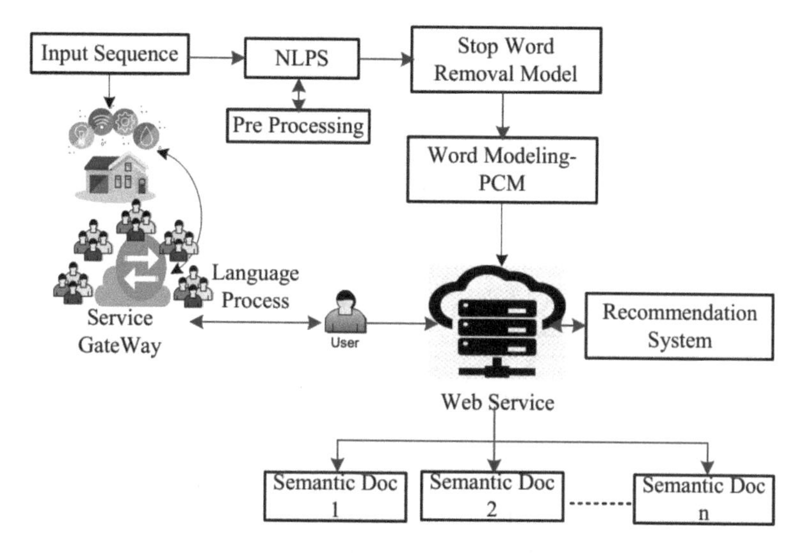

FIGURE 5.3 Block diagram for proposed NLPS model.

5.4.3 Clustering Model

Possibilistic C-means (PCM) clustering algorithm cluster the two prerequisites and competitor administrations, which can viably decrease search space, and this clustering strategy to distinguish the NLPS is dependent on the stop word expulsion process. The imaginative methodology incredibly plays out the movement of simultaneously producing participations and conceivable outcomes. What's more, the novel PCM strategy unbelievably settles the clamor affectability insufficiency of the fuzzy C-means (FCM) clustering altogether overpowering the correspondent clusters challenge of the PCM and eradicates the column entirety parameters of the PCM [19]. The significant state of clustering model is shown in equation (5.6).

$$PCM_{ik} = \frac{1}{\sum_{c}^{j=1}\left(\frac{\|p_k - e_i\|}{\|p_k - e_j\|}\right)^{\frac{2}{m-1}}} \tag{5.6}$$

$$e_1 = \frac{\sum_{n}^{i=1}(PCM_{ik})^m x_k}{\sum_{n}^{i=1} PCM_{ik}} \tag{5.7}$$

This investigation work, the enrollment capacity, and centroid are employed which are adequately furnished in the condition of the NLPS model. Consider the client request $PCM = (pcm_1, pcm_2, \ldots, pcm_n)$ and, create an *m* number of clusters $CH = (ch_1, ch_2, \ldots, ch_m)$. For clustering proposed, PCM limit the target work which is yielded (equation (5.8))

$$\text{Proposed} - PCM(A,B,C;PCM) = \sum_{k=1}^{n}\sum_{i=1}^{c}(aA_{ik}^{m} + BT_{ik}^{n}) \times \|PCM_k - e_i\|_A^2 + \sum_{i=1}^{c}\gamma_i \sum_{k=1}^{n}(1 - T_{ik})^n \tag{5.8}$$

Subject to the parameters $\Sigma_{c}^{i=1} M_{ik} = 1\ \forall k$ and $0 \leq M_{iK}, T_{ik} \leq 1$. Here $a > 0, b > 0, m > 1$ and $\eta > 1$. In equation (5.8), $\gamma_i > 0$ is the client-specified constant. The fixed *a and b* defines the relative importance of ambiguous participation and traditional qualities. From the goal function, A_{ik} is a participation function that is derived from the FCM. The participation function A_{ik} can be calculated as pursues. The cluster center e_i of PFCM is can be calculated as follows:

$$e_i = \frac{\sum_{k=1}^{n} \left(aA_{ik}^{m} + bt_{ik}^{\eta} \right) X_K}{\sum_{k=1}^{n} \left(aA_{ik}^{m} + bt_{ik}^{\eta} \right)}, 1 \leq i \leq c. \tag{5.9}$$

The clustering procedure has proceeded on the *k*-number of emphasis. After the clustering procedure, the client request is gathered into *m* number of clusters. From the methodology of the group-based implanting model, sentence portrayal by language displaying is removed, and scores are calculated which can diminish false positives and arrive at better. This system can be utilized as a stage for any machines that require condition-based applications with no web association. The system will be useful for typical clients likewise and physically incapacitated clients also, as it basically requires voice direction of NLPS. Substances are space explicit data separated from the expression that maps the regular language expressions to their authoritative expressions to comprehend the intent.

5.5 RESULTS AND ANALYSIS

Our proposed NLPS model was executed in JAVA with Netbeans programming windows machine having Intel Core i3 processor with speed 2.10 GHz and 4 GB RAM. This proposed model contrasted and other ordinary strategies with various estimates like precision, review, and F measure. A few methodologies use either NLP or a blend of NLP and machine learning algorithms to deduce details from the normal language components, such as code remarks, depictions, and formal necessities reports.

Figures 5.4 and 5.5 show the cloud-based IoT model for NLPS framework, the login page after a particular interim by sending consequently sensor status get demand by utilizing explicit URIs. At whatever point the smart home door webserver got demand, it gets the sensor status and sends a reaction message to the customer program and furthermore the information is put away in database in the sensors. For clustering model, PCM clustered the, given to the stop word expulsion process for word installing demonstrating record for smart home correspondence through IoT sensors. If chose 30 documents of services means, the top 20 depictions of services too, top-20 rankings are misleadingly removed least in the wake of dissecting and enlarging them, which is considered as the pattern to assess the rankings of service proposal.

Tables 5.1 and 5.2 represent proposed word execution outcomes for incessant stop word evacuation and word demonstrating the implanting process. In this outcomes assessment model, genuine positives and genuine negatives are significant, it's called a perplexity matric table, in given setting because of word sense disambiguation to the savvy home

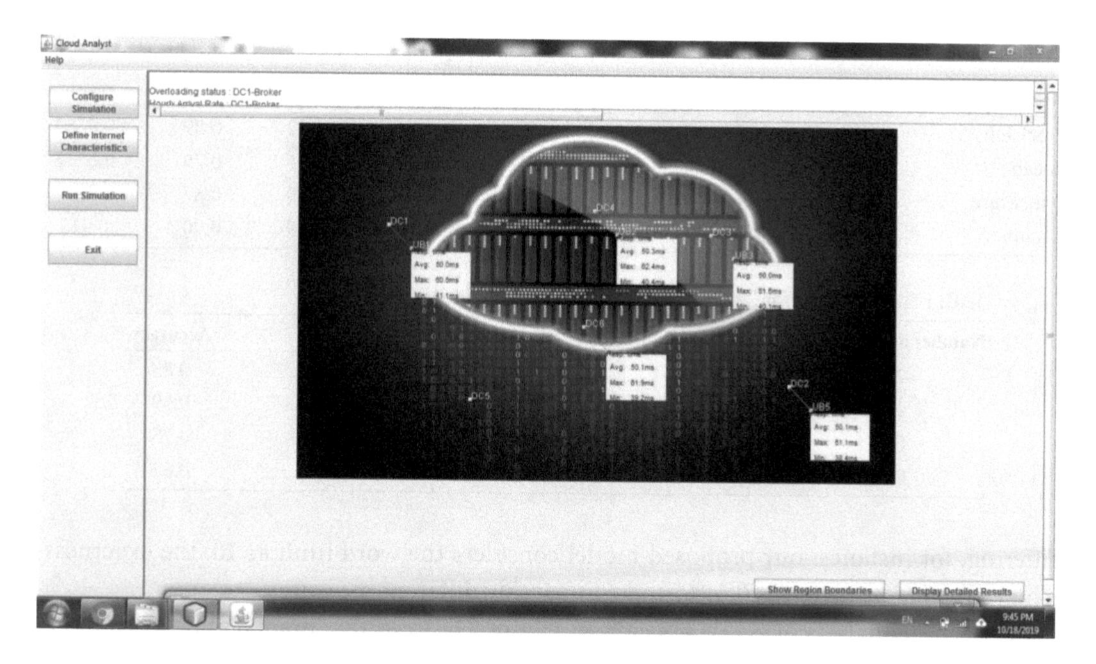

FIGURE 5.4 Smart home NLPS communication process in cloud.

correspondence. Another significant wellspring of false positives is off-base parsing of sentences by NLPS pipeline. Here it uses shallow parsing, where an off-base reliance parsing causes a mistaken development of semantic connection triple. From the experiments conducted the accuracy, exactness, review, and F-measure are 0.93%, 0.85%, 0.8522, and 0.842%, respectively. In the event that emphases fluctuate, the exhibition additionally

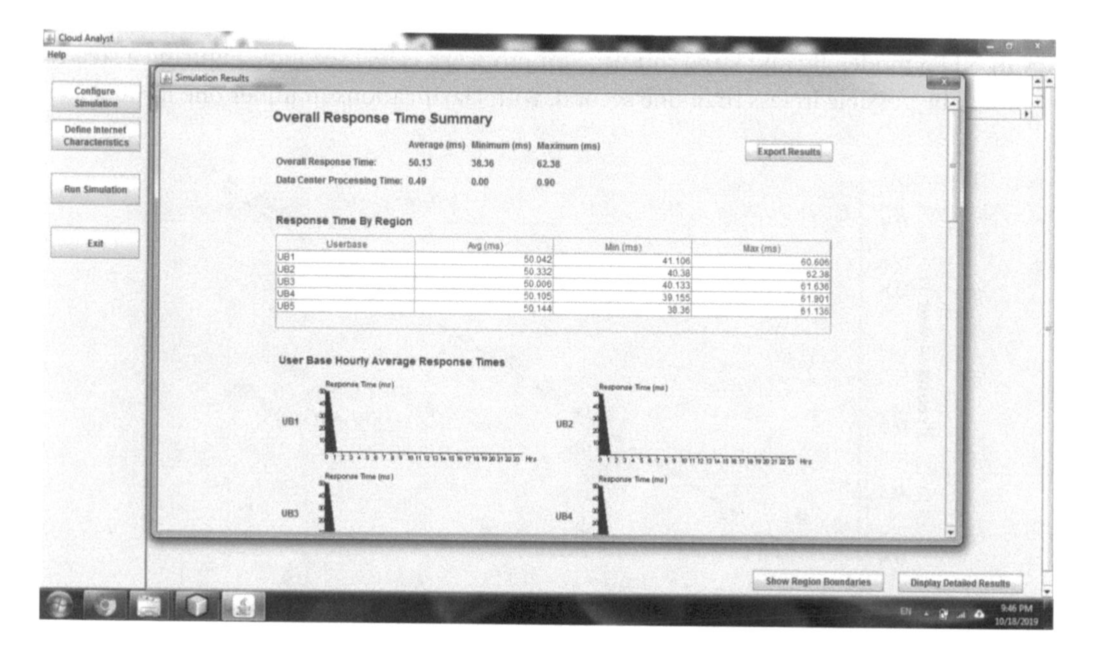

Userbase	Avg (ms)	Min (ms)	Max (ms)
UB1	50.042	41.106	60.606
UB2	50.332	40.38	62.38
UB3	50.006	40.133	61.636
UB4	50.105	39.155	61.901
UB5	50.144	38.36	61.136

FIGURE 5.5 Output user window for NLPS communication model.

TABLE 5.1 Performance Evaluation Metrics for Proposed Smart Home NLPS

Words Count	8	16	24	32	40	48	56	Average
Precision	0.92	0.89	0.86	0.93	0.78	0.92	0.89	0.8783
Recall	0.99	0.86	0.86	0.97	0.89	0.86	0.79	0.8717
F-measure	0.82	0.83	0.79	0.95	0.96	0.89	0.87	0.8817
Accuracy	0.992	0.93	0.89	0.945	0.944	0.89	0.90	0.915

TABLE 5.2 Results for Stop words Removal: Word Modeling

Number of Clusters	Precision	Recall	F-measure	Accuracy
2	0.65	0.75	0.75	0.83
4	0.76	0.76	0.76	0.69
6	0.89	0.89	0.69	0.75
8	0.69	0.78	0.82	0.84

differing, for instance, our proposed model considers the word limit as 10, the exactness achieved is 0.65%, review is 0.61%, and comparably different measures.

Correlation analysis of NLPS appears in Figure 5.6; from this graphical portrayal, the objective reports are considered to the stop word evacuation and language handling in a smart home. The relationship among objective and anticipated NLPS in the smart home correspondence process is cluster-dependent on clustering technique. The link value R^2, the maximum value is 0.956 and the optimum semantic subspace, while at the same time improving the similarities between the records in the nearby fixes and restricting the interaction between the archives beyond these patches. The stop words which ought to be expelled are given legitimately. Need to dispose of those stop words for finding such similitude between records. At that point the run time (second) appears in Figure 5.7. Even if the cycle shifts, the time additionally changes yet the proposed model gets most extreme accuracy in NLPS model. By and large run time for our work is 989 seconds. Figuring a steady set should be possible in less than one second, with clarifications in under one minute.

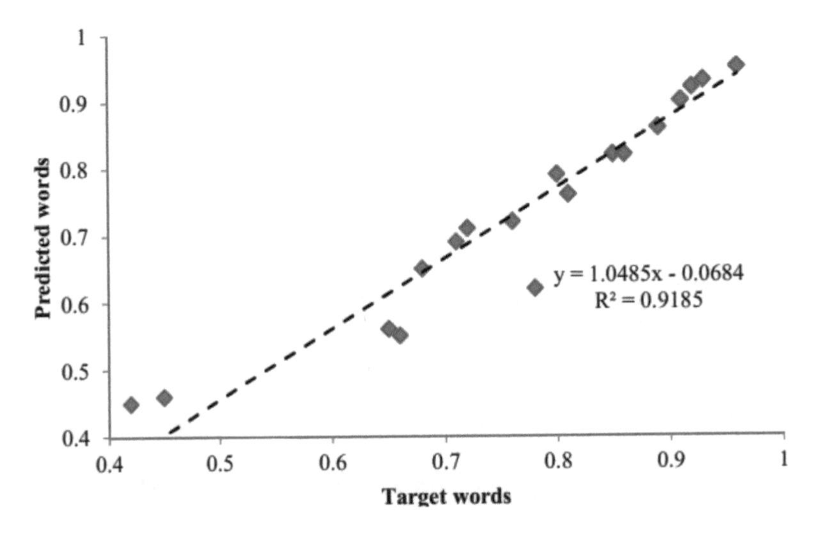

FIGURE 5.6 Correlation analysis of proposed NLPS.

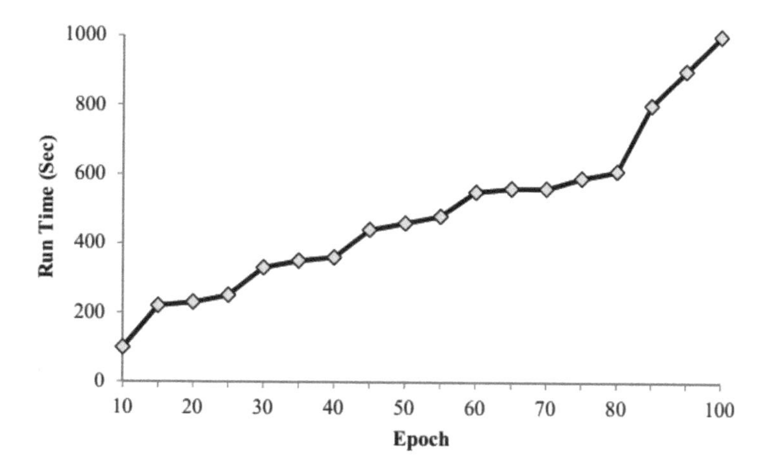

FIGURE 5.7 Execution time for analysis.

5.5.1 Comparative Analysis of NLPS

Smart home correspondence appears in Figure 5.8. Here, three methods are utilized for comparison, that is, PCM, FCM, and K-means clustering model. The most extreme accuracy of PCM is 0.89%, it's contrasted with the fluffy model the thing that matters is 5.56–8.85% recall proportion of various clustering procedure for NLPS. This is a direct result of the fuzzy. The exhibition of the proposed methodology is investigated as far as F measure. Here, additionally, our proposed methodology accomplishes better outcomes. The greatest F proportion is 0.96 which is highly contrasted with different algorithms. This is a direct result of K-means clustering. The PCM beat the troubles present in the Fuzzy and proposed model. In this way, in this chapter PCM-based technique accomplishes better outcomes. From the outcome, we unmistakably comprehend our proposed strategy accomplish a superior outcome contrast with another strategy due to clustering procedure.

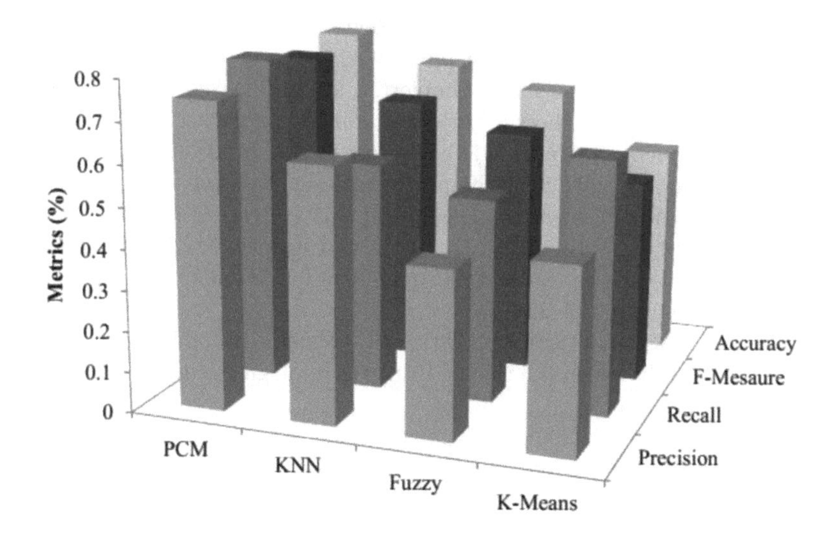

FIGURE 5.8 Comparative analysis with conventional techniques.

5.6 CONCLUSION

Despite these promising outcomes, some energizing challenges remain, particularly so as to scale-up this model to necessities with progressively complex semantics. In this chapter, we have examined the NLPS in smart home applications or interchanges by stop word expulsion and word demonstrating strategy by PCM. Embedding as its center, our methodology can be effectively prepared on marked information and fundamentally outflanks traditional work. That is the reason the explained strategy gives the possibility to make regular language interfaces to complex programming. This technique enables us to create natural language interfaces for activity-based applications with help of client directions from easy to long complex guidelines. Most extreme execution estimates such as precision, review, F measure are 0.865%, 0.75%, and 0.92%, respectively.

These voices and the order prepared by natural language handling encourage clients to create a superior relationship with innovation and encourage them to make more use of it in smart home applications with IoT sensors.

REFERENCES

1. Caivano, D., Fogli, D., Lanzilotti, R., Piccinno, A. and Cassano, F., 2018. Supporting end-users to control their smart home: design implications from a literature review and an empirical investigation. *Journal of Systems and Software*, 144, pp. 295–313.
2. Zaidan, A.A. and Zaidan, B.B., 2018. A review of intelligent process for smart home applications based on IoT: coherent taxonomy, motivation, open challenges, and recommendations. *Artificial Intelligence Review*, 53, pp. 141–165.
3. Rani, P.J., Bakthakumar, J., Kumaar, B.P., Kumaar, U.P. and Kumar, S., 2017. Voice controlled home automation system using natural language processing (NLP) and internet of things (IoT). In *2017 Third International Conference on Science Technology Engineering & Management (ICONSTEM)* (pp. 368–373). IEEE.
4. Iqbal, R., Lee, J. and Hall, J., 2018. A cloud middleware enabling natural speech analysis for IoT policy enforcement in smart home environments. In *2018 IEEE International Congress on Internet of Things (ICIOT)* (pp. 184–187). IEEE.
5. Guo, X., Shen, Z., Zhang, Y. and Wu, T., 2019. Review on the application of artificial intelligence in smart homes. *Smart Cities*, 2(3), pp. 402–420.
6. Fernández, M., Montalvá, J.B., Cabrera-Umpierrez, M.F. and Arredondo, M.T., 2009. Natural language interface for smart homes. In *International Conference on Universal Access in Human-Computer Interaction* (pp. 49–56). Springer, Berlin, Heidelberg.
7. Badlani, A. and Bhanot, S., 2011. Smart home system design based on artificial neural networks. In *Proceedings of the World Congress on Engineering and Computer Science* (Vol. 1, pp. 146–164).
8. Bashir, A.M., Hassan, A., Rosman, B., Duma, D. and Ahmed, M., 2018. Implementation of A neural natural language understanding component for Arabic dialogue systems. *Procedia Computer Science*, 142, pp. 222–229.
9. Hui, T.K., Sherratt, R.S. and Sánchez, D.D., 2017. Major requirements for building Smart homes in smart cities based on Internet of Things technologies. *Future Generation Computer Systems*, 76, pp. 358–369.
10. Farghaly, A. and Shaalan, K., 2009. Arabic natural language processing: challenges and solutions. *ACM Transactions on Asian Language Information Processing (TALIP)*, 8(4), p. 14.
11. Moubaiddin, A., Shalbak, O., Hammo, B. and Obeid, N., 2015. Arabic dialogue system for hotel reservation based on natural language processing techniques. *Computación y Sistemas*, 19(1), pp. 119–134.

12. Mishakova, A., Portet, F., Desot, T. and Vacher, M., 2019. Learning natural language understanding systems from unaligned labels for voice command in smart homes. In *2019 IEEE International Conference on Pervasive Computing and Communications Workshops (PerCom Workshops)* (pp. 832–837). IEEE.
13. Alexakis, G., Panagiotakis, S., Fragkakis, A., Markakis, E. and Vassilakis, K., 2019. Control of smart home operations using natural language processing, voice recognition and IoT technologies in a multi-tier architecture. *Designs*, 3(3), p. 32.
14. Noguera-Arnaldos, J.Á., Paredes-Valverde, M.A., Salas-Zárate, M.P., Rodríguez-García, M.Á., Valencia-García, R. and Ochoa, J.L., 2017. im4Things: an ontology-based natural language interface for controlling devices in the Internet of Things. In *Current Trends on Knowledge-Based Systems* (pp. 3–22). Springer, Cham.
15. Jivani, F.D., Malvankar, M. and Shankarmani, R., 2018. A voice controlled smart home solution with a centralized management framework implemented using AI and NLP. In *2018 International Conference on Current Trends towards Converging Technologies (ICCTCT)* (pp. 1–5). IEEE.
16. Qin, P. and Guo, J., 2020. A novel machine natural language mediation for semantic document exchange in smart city. *Future Generation Computer Systems*, 102, pp. 810–826.
17. Akerkar, R. and Joshi, M., 2008. Natural language interface using shallow parsing. *IJCSA*, 5(3), pp. 70–90.
18. Raulji, J.K. and Saini, J.R., 2016. Stop-word removal algorithm and its implementation for Sanskrit language. *International Journal of Computer Applications*, 150(2), pp. 15–17.
19. Yu, H., Fan, J. and Lan, R., 2019. Suppressed possibilistic c-means clustering algorithm. *Applied Soft Computing*, 80, pp. 845–872.

CHAPTER 6

Metaheuristic-Based Kernel Extreme Learning Machine Model for Disease Diagnosis in Industrial Internet of Things Sensor Networks

S. Dhanasekaran,[1] I. S. Hephzi Punithavathi,[2] P. Duraipandy,[3] A. Sivanesh Kumar,[4] P. Vijayakarthik,[5] S. Rajasekaran,[6] and B. S. Murugan[7]

[1]*Department of Computer Science and Engineering, Kalasalingam University, Srivilliputtur, Tamil Nadu, India*

[2]*Department of Computer Science and Engineering, Sphoorthy Engineering College, Hyderabad, Telangana, India*

[3]*Department of Electrical and Electronics Engineering, J.B. Institute of Engineering and Technology, Hyderabad, Telangana, India*

[4]*Department of Computer Science and Engineering, Saveetha School of Engineering, Saveetha Institute of Medical and Technical Sciences, Chennai, Tamil Nadu, India*

[5]*Department of Information Science and Engineering, Sir M. Visvesvaraya Institute of Technology, Bangalore, Karnataka, India*

[6]*Department of EEE, PSN College of Engineering, Anna University, Tirunelveli, Tamil Nadu, India*

[7]*Department of CSE, Kalasalingam Academy of Research and Education, Srivilliputtur, Tamil Nadu, India*

CONTENTS

6.1 INTRODUCTION

Nowadays, enhanced utilization of intelligent tools and communication applications in medical observance and the control on actions of medical employees such as doctors, nurses, and hospital managers, patients, and healthcare manufacturers have been witnessed. Based on Gartner and Forbes, it is estimated that the Internet of Things (IoT) has contributed maximum financial cost for the global economy and minimum amount for IoT-based healthcare production [1]. According to these estimates, it can be observed that the Health IIoT is mostly an important player in the Industrial IoTs (IIoTs)-motivated healthcare sector. IIoT has had a remarkable control over several massive and tiny healthcare manufacturing domains. Thus, an enhanced amount of carrying IoT tools, machines, and apps are being utilized to monitor several health-related details such as glucose meter, ECG screens, and blood pressure (BP), etc.

Figure 6.1 shows the industrial IoT healthcare ecosystem.

At present, Health IIoT is in its beginning steps with the consideration for designing and utilization. But IoT-based results are currently showing an extraordinary effect and carving out a developing market. IIoT has the possibility of saving maximum people annually in the United States by eliminating the mortality rate due to the limited medical facilities. It assures patient health and protection by managing vital patient data and synchronizes compared resources (e.g., healthcare staff, services, carrying smart tools for capturing concurrent patient information like critical signs, and patient-compared electronic data) instantaneously with interrelated tools and sensors. It reveals that IoT in the healthcare production makes possible optimal care with costs, diminished direct patient-healthcare staff interface, and ubiquitous allowance for quality care.

Mohammed et al. [2] supposed to design a remote patient observing model utilizing web services and cloud computing (CC). Protected and high-quality healthcare diagnosis is of paramount significance to patients. Consequently, healthcare information protection and patients' privacy are the significant problems which caused a huge impact on the prospective achievement of Health IIoT. The mainly revolutionary possible function is healthcare observing, as patient healthcare information is gathered from a count of sensors in a network and distributed through healthcare trains to estimate patient care.

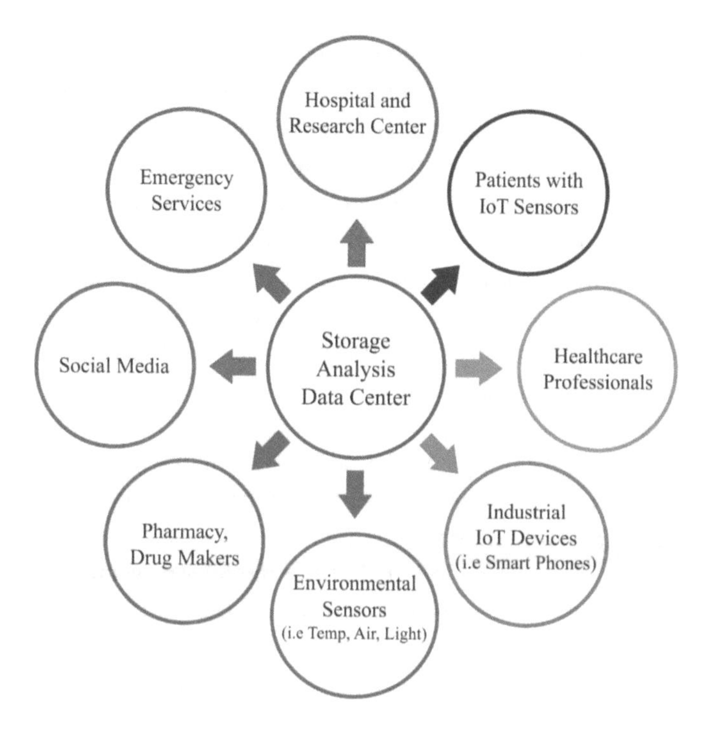

FIGURE 6.1 Industrial IoT healthcare ecosystem.

A further widespread investigation of IoT in healthcare functions is established in [3]. The IoT-enabled healthcare functions containing IoT-driven ECG observations are explained in the subsequent studies.

Li et al. [4] proposed a health observance as a platform to observe ECG utilizing an adaptive learning examination method for detecting anomalies. Mohammed et al. supposed a remote patient observing model utilizing web services and CC. Particularly, it has planned an Android function to ECG information observing as well as investigation. Information is examined with third-party software when required; but, it is not an opportunity for the cloud server for extracting features. It classifies the signal for assisting the health trains at the time the signal is obtained. In the presented structure, the cloud server removes features and classifies the signal. In order to examine this signal, the decision from the cloud is sent to the healthcare trains for facilitating optimal patient care.

Hassanalieragh et al. [5] explained the options and challenges of health observing and organization utilizing IoT. Several challenges contain slowdown the procedure, managing big data, the occurrence of extra heterogeneous information, and data integrity. During the presented structure, the ECG signal is watermarked on the client-side previous to broadcasting with the Internet for authenticating over some attacks. The data procedure is also shared among the client-side as well as the cloud-side for making the entire method quicker. Jara et al. [6] proposed a remote observing structure utilizing IoT with a presented protocol, known as YOAPY, for creating a safety and scalable fusion of multi-modal sensors for recording critical signs. A cloud-based speech and face identification structure were extended for monitoring a patient's condition distantly.

Xu et al. [7] offered a ubiquitous data allowing technique in an IoT-based model to emergency medicinal conditions. It has presented a semantic data system for storing information and a resource-based information access technique for gaining management of the data ubiquitously. It is helpful for assisting decision-generating in emergency medicinal conditions. Zhang et al. [8] established a structural design of mobile healthcare networks, which includes privacy-maintaining information gathered and safety broadcasting. The privacy-preserving information gathered is obtained utilizing cryptography through secret as well as private keys. A secure broadcast is obtained utilizing attribute-based encryption where only an allowed user can access the information. This technique is usually useful; but, the major issue is calculation difficulty.

Granados et al. [9] presented web-enabled gateways to IoT-based eHealth through the opportunity to wired or wireless services. For taking benefit of wired gateways with respect to power-efficiency and minimum cost, it has utilized the wired gateways in a tiny room or structure, as progresses are limited. Radio frequency identification (RFID)-based eHealthcare modes were presented. In Amendola et al. [10], a model is projected which captures the patient's surrounding situations such as temperature and humidity by RFID and broadcasts them for the cloud to a further, brief comprehension of ambient situations. Catarinucci et al. [11] projected an IoT-aware structural design for monitoring and assessing a patient's condition automatically by combining ultra-high-frequency RFID functions.

Sawand et al. [12] recognized three kinds of threats in an eHealthcare observing model. This is an identity threat, as the identity of the patient is lost or stolen, allowing threat, as an intruder is allowing the model illegally, and disclosure threat, as secret medicinal information are opened using malware or record distributing devices intentionally or unintentionally. It suggested several results for this threat, containing biometric cryptography and an advanced signal procedure method; but, it does not execute the results in their study. A patient with IoT devices and sensors can be used to training a smart healthcare diagnosis model for drug manufacturing and pharmacies (e.g., smart pills, and medicines).

In Chen et al. [13], an emotion-aware or affective mobile computing structures have been projected, and the authors examined the structural design like "emotion-aware mobile cloud" (EMC) for mobile computing. Chen et al. [14] presented another structure, affective interaction through wearable computing and cloud technology (AIWAC). Currently, Hu et al. [15] established the Healthcare IoTs (Health-IoT), trying to bridge among intelligence health observing and emotional care of the patients. So far, it has established no understanding learn on cloud-assisted IIoT-driven ECG observing, as i) an ECG signal is watermarked on the client-side previous to broadcasting with the Internet for the cloud and ii) a cloud server removes features and classifiers the signal for assisting healthcare trains in giving feature patient care.

This paper develops a new Spider Monkey Optimization (SMO)-based Kernel Extreme Learning Machine (KELM) model for disease diagnosis in IIoT environment. The proposed model at first gathers the medical data from heart disease patients utilizing IoT devices. Then, the IoT devices execute data compression by means of Deflate algorithm to diminish the quantity of data being transmitted to a cloud. Concurrently, the cloud server decompresses the data and executes the SMO-KELM model for HD diagnosis. In SMO-KELM model, the parameter tuning of KELM model takes place using SMO algorithm, which depends upon the nature of spider monkeys. Lastly, the alarm module is used to generate an alert in case of the existence of HD. An extensive experimentation will be done to verify the goodness of the presented model.

6.2 PROPOSED METHODOLOGY

Figure 6.2 implies the whole architecture of processes involved in SMO-KELM model, which incorporates data collection, data compression, disease diagnosis, and alert system. At the initial stage, the IoT devices are fixed to patient's body and collect the patient data, and it is compressed using Deflate method. Once the patient data has been compressed, it is transmitted to the cloud server via wireless technologies. The cloud server performs decompression process and reconstructs the data proficiently. Then, the SMO-KELM model is executed to detect the presence of disease. Finally, an alarm will be raised in case of the occurrence of the diseases in real time to alert doctors, ambulance, and hospitals.

6.2.1 Deflate-Based Compression Model

Deflate is referred as lossless compression technique which has been extremely utilized over an extended time duration because of its maximum speed and optimal compression effectiveness [16]. Several techniques such as GZIP, ZLIB, ZIP, and PKZIP depend on the

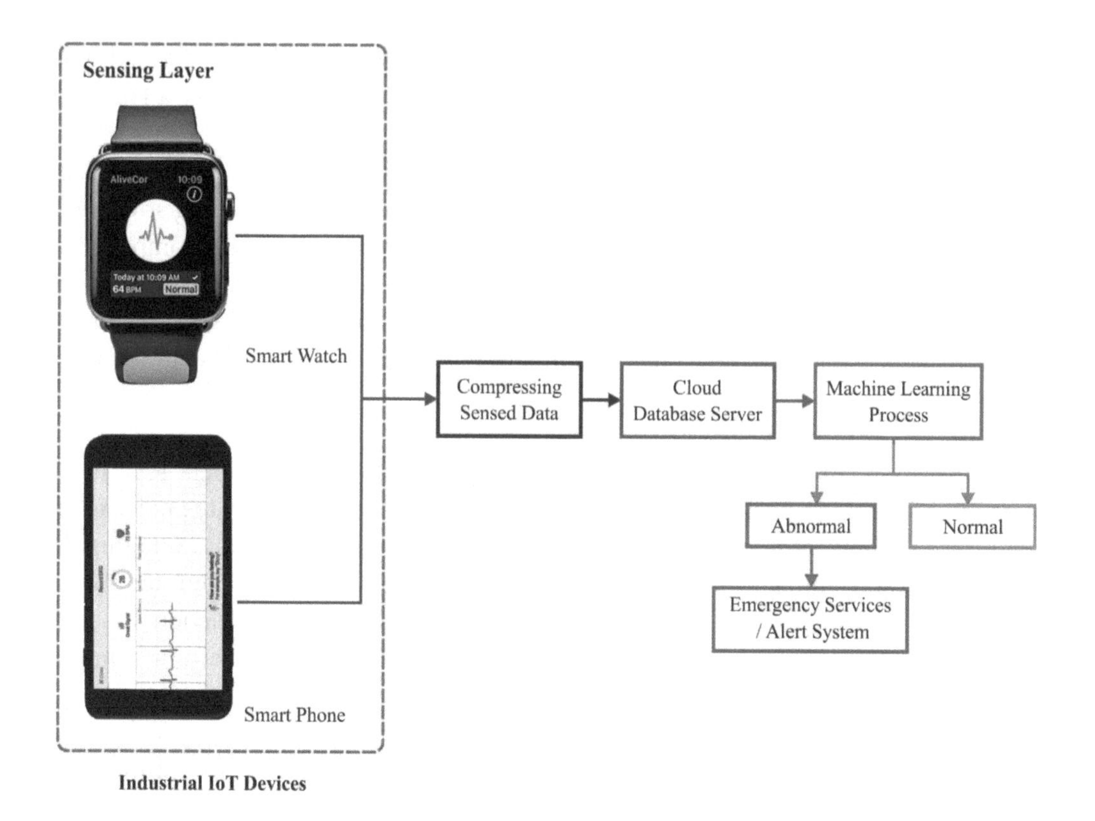

FIGURE 6.2 Block diagram of proposed model.

Deflate compression technique. These techniques contain LZ77 method and Huffman coding. The original information undergoes initial compression utilizing the LZ77 technique and after that the data is even reduced by the Huffman technique.

6.2.1.1 LZ77 Encoding

The LZ77 technique is a dictionary-centric compression technique. It has developed a dictionary with adjacent strings. For input string, dictionary is identified. If there is an equivalent initiate, the projected procedure string is returned with a distance as well as length of string recorded in a dictionary. Since the string is not mapped, the primary incidence of string is remained the same in the dictionary application. It returns a group of data which contains comparative position (corresponding distance), length of the corresponding string (corresponding length), and a flag denoting a chunk of information is encoded (marker). Assume that the corresponding length and distance is indicated as length as well as distance.

6.2.1.2 Huffman Coding

This coding is defined as a type of entropy coding that limits data by assigning shorter bits for happening symbols. It is composed of two technologies like Huffman code generation as well as Huffman encoding models. The Huffman tree (HT) is based on the frequency of symbols applied in the data compression. At the initial stage, two symbols with lower frequency have been selected. Here, two leaf nodes have been developed by selected symbols which are combined to develop a new node. The creation strategy is prominently applied for all symbols. The main aim of HT is that the tree assigns shorter codes for repeated symbols and longer codes to lower repeated symbols. Such computations produce a Huffman code for all symbols in the LZ77 encoding stream.

The encoded model reduces the LZ77 encoding stream utilizing the Huffman code tables build initially. During the model, all the symbols in the LZ77 encoding stream are returned by a corresponding Huffman code. The LZ77 encoding stream is comprised of variable length data elements. Additionally, Huffman encoding changes the elements of data using variable-element codes. Thus, encoded procedure analyzes the stream components, and alters using Huffman codes, and finally, it is fixed with resultant data stream.

6.2.2 SMO-KELM-Based Diagnosis Model

Back propagation (BP) learning technique is defined as a stochastic gradient least mean square technique. A gradient of all iterations is significantly concerned by the noise interfering in the sample. So, it will be essential for utilizing the batch model for averaging the gradient of several samples for obtaining the evaluation of the gradient. But, for a huge count of training samples, these techniques are bound to increase the calculation difficulty of all iterations. The average result ignores the variation among separate training samples, thus diminishing the sensitivity of learning.

KELM is defined as an enhanced technique that joins the kernel function with actual Extreme Learning Machine (ELM). The ELM assures that a network with optimal generalized execution optimally enhances the learning speed of Neural Networks (NN) and evades several issues of gradient descent (GD) training techniques signified with back propagation neural network (BPNN), such as simplicity of being trapped in local optimal, huge rounds, and so on. It is a multi-dimensional ELM model, although it joins the kernel function that non-linearly maps the linear, non-separable mode for the highly dimensional feature space for attaining a linear, separable, and more enhanced rate of accuracy.

ELM is defined as a training method, which comprises single layer feed-forward NNs (SLFNs). The SLFN model is determined as follows:

$$f(x) = h(x)\beta = H\beta \tag{6.1}$$

where x denotes a sample, $f(x)$ implies the outcome of NNs, which is a class vector in classification model; $h(x)$ or H represents the layer feature mapping matrix; β is defined as weight of hidden layer. For ELM approach,

$$\beta = H^T\left(HH^T + \frac{I}{C}\right)^{-1} T \tag{6.2}$$

where T represents a matrix with class flag vectors of the training sample, I implies a unit matrix, and C denotes the regularized attribute.

As the hidden layer has a feature map, $h(x)$ is uncertain, and KELM matrix is computed in the following:

$$\Omega = HH^T : \Omega_{ij} = h(x_i)\cdot h(x_j) = K(x_i, x_j). \tag{6.3}$$

Based on equations (6.2) and (6.3), equation (6.1) is transformed as follows:

$$\begin{aligned} f(x) &= H\beta = HH^T\left(HH^T + \frac{I}{C}\right)^{-1} T \\ &= \begin{bmatrix} K(x, x_1) \\ \vdots \\ K(x, x_N) \end{bmatrix}^T \left(\Omega + \frac{I}{C}\right)^{-1} T. \end{aligned} \tag{6.4}$$

Under the application of Radial Basis Function (RBF), Gaussian kernel function is calculated in the following:

$$K(x, y) = \exp\left(-\frac{\|x - y\|^2}{2\gamma^2}\right), \tag{6.5}$$

Then, a regularized attribute C and a kernel function parameter γ are variables that are essential for proper tuning. The patterns of C and γ are vital factors that affect the execution of KELM classification [17]. However, the variables of KELM should be optimized using SMO technique.

The SMO technique is a metaheuristic technique depends on spider monkey's social performance, accepting the fission as well as fusion of swarm intelligence (SI) approach to forage. Generally, spider monkeys reside in a swarm. A leader is appointed for classifying the responsibility of searching food. Then, a female, being a leader, directs the swarm and provides mutable sets when the food is minimum. The set is based on food accessibility from specific application. The SMO relied model ensures the subsequent essential fundamentals of SI:

- Labor division: A spider monkey separates the exploring process under the development of tiny sets.
- Self-accessibility: The set size can be selected in order to satisfy the food accessibility.

The foraging intelligence performance assists in the creation of an intelligence decision. The foraging behavior is explained with the subsequent stages as given below:

1. The swarm founds food exploration.
2. Calculate distance from food sources.
3. The distance of swarm from the food accomplishes while changing the positions.
4. Again, the distance of swarm from a food source has been processed.

Here, a local leader contributes to find an optimal location from sub-swarms. These locations alter over time based on food accessibility. If the position is not changed by a local leader from a sub-swarm, then it is applicable for all iterations, the subgroup members are self-estimated by shifting freely in various models. Alternatively, global leader involves in finding a good position for all members of the swarm. Also, locations are changed on the basis of food accessibility. At the point of immobility, it selects the swarm into a sub-swarm of reduced sizes. The exceeded stages undergo iteration until the teaching-defined result can be accomplished. Hence, the SMO-based approach has been categorized as a nature-based model which depends upon SI.

6.2.2.1 Key Steps of SMO Algorithm Implementation

SMO is defined as a population-oriented method that approves trial-and-error relied on collaborative iterative models with six stages, such as local leader, local leader learning, local leader decision, global leader, global leader learning, and global leader decision stages [18]. The iterative process of SMO execution is explained in the following sections.

6.2.2.1.1 Initializing the Population SMO distributes the population of P spider monkeys SM_p uniformly where $p = 1, 2, \ldots, p$ and SM_p indicates the pth monkey of a population. Then, monkeys are regarded as M-dimensional vectors, as M determines the entire amount of variables in the issue fields. All SM_p is compared to one feasible result to the provided issue. SMO establishes all SM_p utilizing the subsequent equation (6.6):

$$SM_{pq} = SM_{minq} + UR(0,1) \times \left(SM_{maxq} - SM_{minq}\right) \tag{6.6}$$

where, SM_{pq} defines the qth dimension of pth SM; SM_{minq} and SM_{maxq} are maximum and minimum bounds, respectively, of SM_p in the qth way in which $q = 1, 2, \ldots, M$ UR (0, 1) mimics the random value that is distributed uniformly within [0, 1].

6.2.2.1.2 Local Leader Phase (LLP) In LLP, the SM modifies the recent position using previous incidences of local leader as well as local set candidates. The SM location has been upgraded with new location, when the novel position is composed of a fitness value which is qualified than existing places. The expression of location update for pth SM of lth local set can be determined as:

$$SMnew_{pq} = SM_{pq} + UR(0,1) \times \left(GL_{lq} - SM_{pq}\right) + UR(-1,1) \times \left(SM_{rq} - SM_{pq}\right) \tag{6.7}$$

where GLq shows the position of global leader in qth dimension and $(q = 1, 2, 3, \ldots, M)$ defines the arbitrarily selected index.

In this point, the fitness of SM has been applied for estimating the probability prb_p. Based on the possibility value, SM_p position has been extended in these approaches. Best position is composed of access for improved number of possibilities to deploying better one. The probability computation can be estimated using the given expression:

$$prb_p = \frac{fn_p}{\sum_{p=1}^{N} fn_p} \tag{6.8}$$

where fn_p showcases the fitness score of pth SM. In addition, fitness of new place of SMs is determined and relevant to previous location. The place with best fitness value has been approved.

6.2.2.1.3 Global Leader Learning (GLL) Phase The greedy selection methods were implemented for extending the location of GL. This objective has been enhanced using SM position, followed by optimal fitness measure in the population. The best places were assigned for GL. In absence of additional updates, the value of 1 is included to Global Limit Count.

6.2.2.1.4 Local Leader Learning (LLL) Phase In this phase, greedy selection models have been applied in local set for the purpose of upgrading LL position. The LL location can be maximized using SM place under the employment of better fitness value in certain local group. The best positions are assigned for LL. When there is no supplement updates, the measure of 1 is added to Local Limit Count.

6.2.2.1.5 Local Leader Decision (LLD) Phase When LL is not upgraded with the position in applicable Local Leader Limit, all candidates of local group adjust the places in a random manner as given in step 1, it can also be accomplished using existing data from GL and LL on the basis of pr as given in equation (6.9).

$$SMnew_{pq} = SM_{pq} + UR(0,1) \times \left(GL_{lq} - SM_{pq}\right) + UR(0,1) \times \left(SM_{rq} - LL_{pq}\right) \tag{6.9}$$

6.2.2.1.6 Global Leader Decision (GLD) Phase If GL is not improved for its position till reaching Global Leader Limit, the population is divided into minimum sets, as per solution of GL. This procedure is repeated until declared with maximum groups (MG). For each iteration, an LL has been selected for novelty developed group. While bigger value is allocated for produced and GL does not maximize the location, and GL selects the way of combining the entire sets into an individual set.

6.3 EXPERIMENTAL RESULTS AND VALIDATION

The efficiency of the Deflate-based compression (D-comp) method and SMO-KELM method has been tested under diverse perspectives. Table 6.1 and Figure 6.3 provide a comparative analysis of the compression efficiency of the D-comp model. The table values indicated that the D-comp method compresses the original size to minimal size compared to LZW model. At 2000 instances, the D-comp method has compressed the original file of 19,820 bytes into 12,395 bytes, whereas the LZW model has compressed it into 15,639 bytes only. At 4000 instances, the D-comp approach has compressed the original file of 36,391 bytes into 23,756 bytes while the LZW method has compressed it into 16,492 bytes only. At 6000 instances, the D-comp model has compressed the original file of 55,689 bytes into 46,932 bytes, but the LZW approach has compressed it into 35,872 bytes only. At 8000 instances, the D-comp model has compressed the original file of 74,310 bytes into 58,429 bytes while the LZW method has compressed it into 45,396 bytes only. At 10,000 instances, the D-comp model has compressed the original file of 92,096 bytes into 78,424 bytes, but the LZW method has compressed it into 65,839 bytes only.

Table 6.2 and Figure 6.4 analyze the SS performance of the D-comp and LZW models under varying number of instances. The figure indicated that the presented D-comp model has attained maximum SS over the LZW method. At the 2000 instances, the D-comp method has attained a maximum SS of 37.46%, whereas the LZW model has offered a lower SS of 21.09%. At the 4000 instances, the D-comp model has obtained a highest SS of 54.68% while the LZW method has offered a lower SS of 34.72%. At the 6000 instances,

TABLE 6.1 Analysis of Compressed File Size vs. Original Size on Proposed D-Comp Method

Number of Instances	Original Size (Bytes)	LZW (Bytes)	D-Comp (Bytes)
2000	19,820	15,639	12,395
4000	36,391	23,756	16,492
6000	55,689	46,932	35,872
8000	74,310	58,429	45,396
10,000	92,096	78,424	65,839

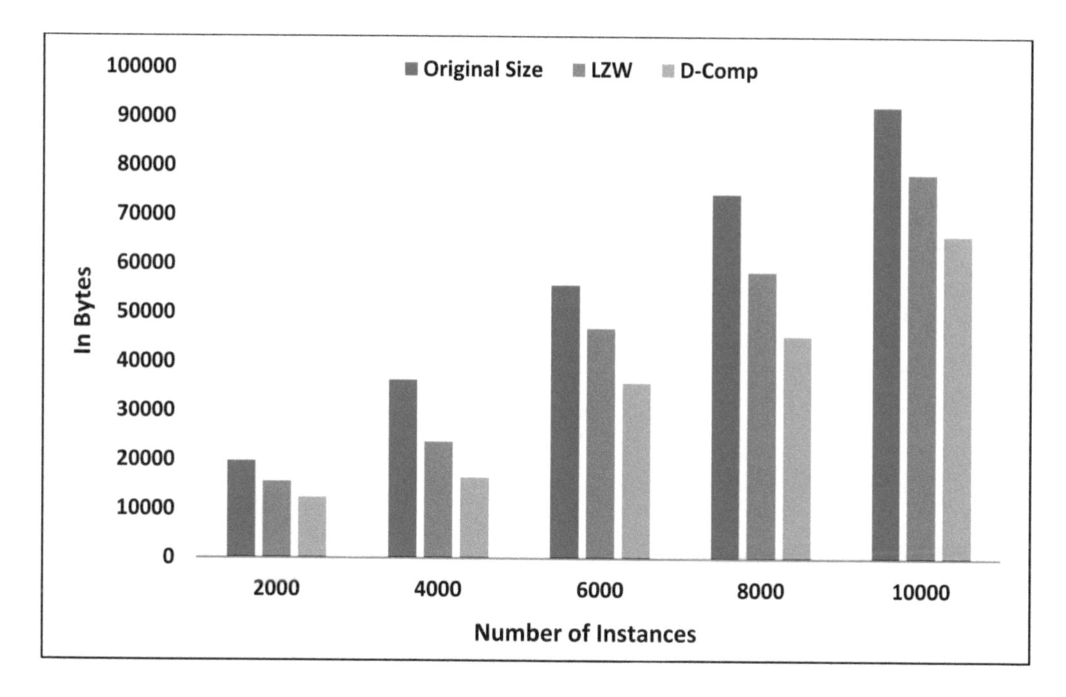

FIGURE 6.3 Comparative analysis of compression efficiency D-comp model.

the D-comp approach has achieved a maximum SS of 35.59%, but the LZW method has offered a minimum SS of 15.72%. At the 8000 instances, the D-comp methodology has obtained a highest SS of 38.91%, but the LZW approach has offered a lower SS of 21.37%. At the 10,000 instances, the D-comp approach has obtained a maximum SS of 28.51% while the LZW model has offered a minimum SS of 14.85%.

Figure 6.5 depicts the analysis of the results offered by the SMO-KELM model in terms of sensitivity. The figure stated the support vector machine (SVM) and Naïve Bayes (NB) models have exhibited ineffective classification performance by attaining minimal sensitivity values. Besides, the k-nearest neighbor (KNN) model has achieved slightly higher sensitivity value over the compared methods. At the same time, the decision tree (DT) model has exhibited competitive results compared to other methods. But the presented SMO-KELM model has surpassed all the compared methods and obtained maximum sensitivity under all the instances. For instance, under the instance count of 2000, the

TABLE 6.2 Analysis of Space Savings on Compressed File Size vs Original Size on Proposed D-Comp Method

	Space Savings	
Number of Instances	LZW	D-Comp
2000	21.09	37.46
4000	34.72	54.68
6000	15.72	35.59
8000	21.37	38.91
10,000	14.85	28.51

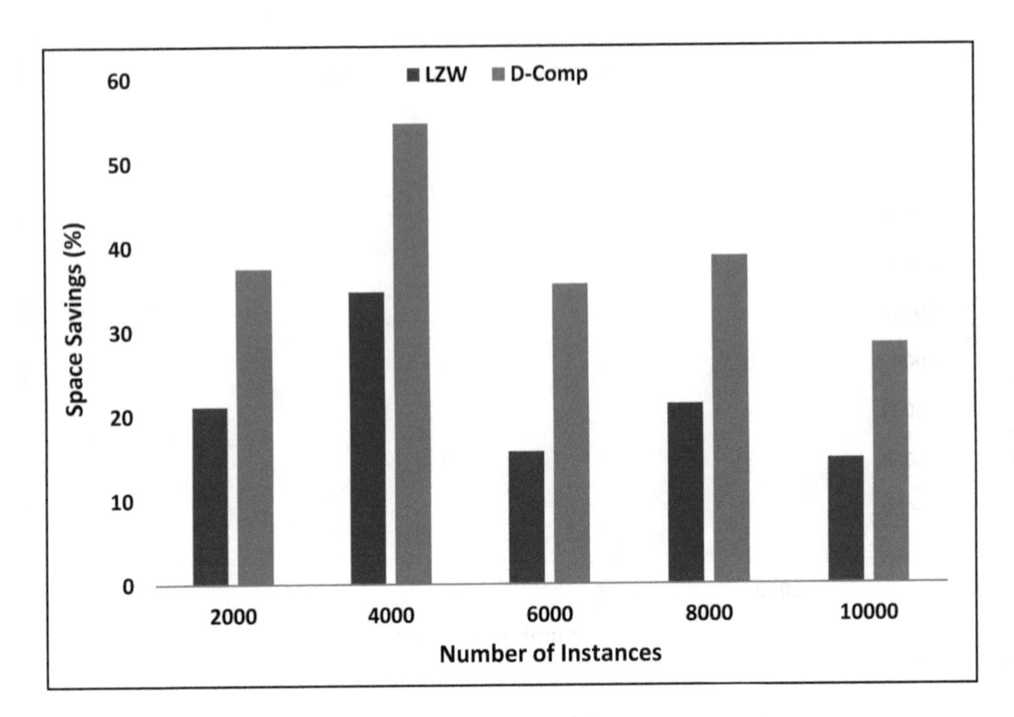

FIGURE 6.4 Space saving analysis of D-comp and LZW models.

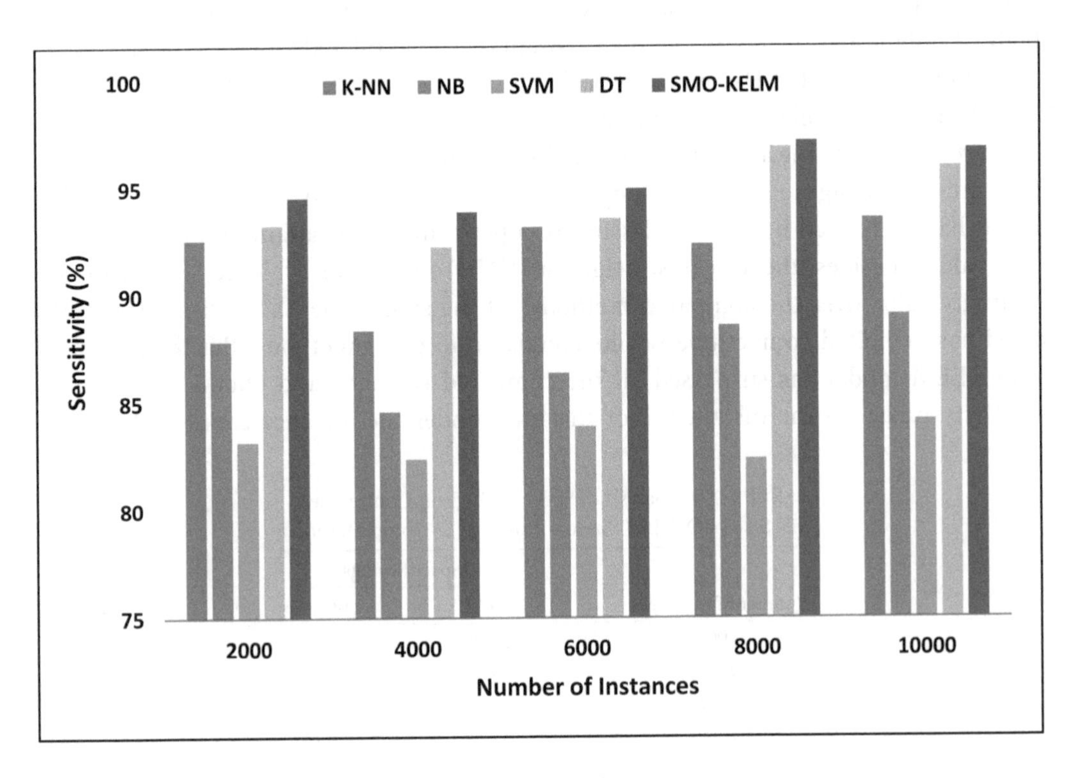

FIGURE 6.5 Sensitivity analysis of SMO-KELM model.

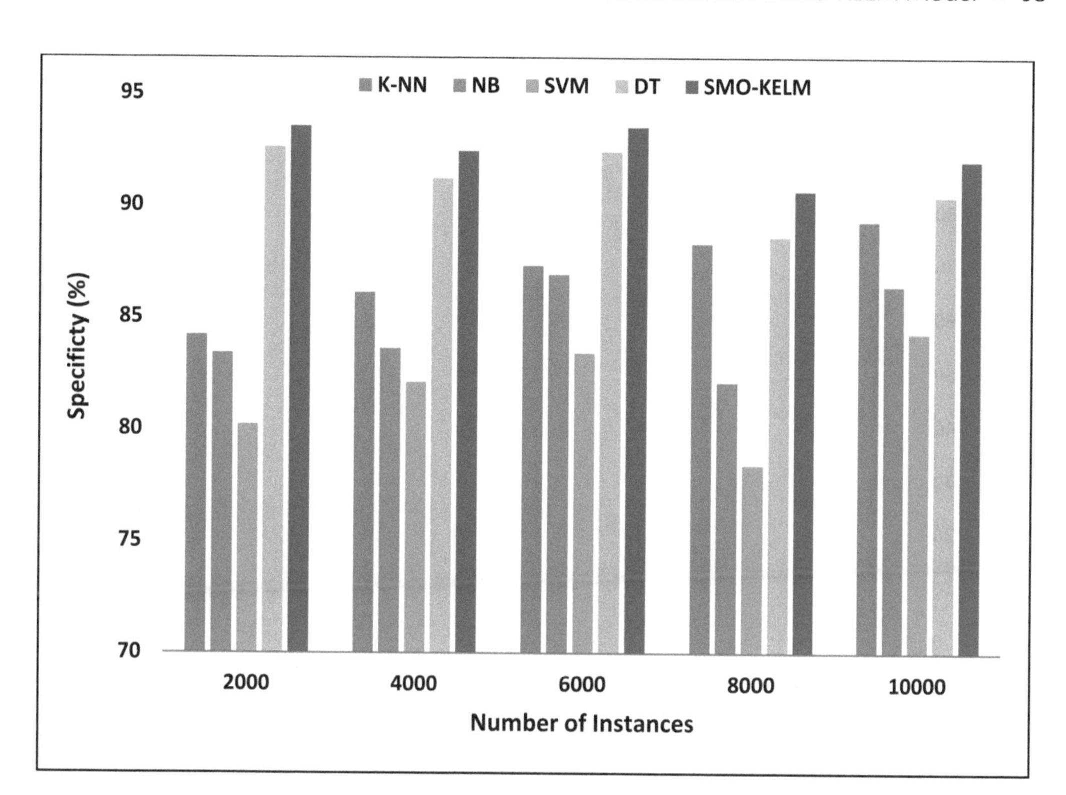

FIGURE 6.6 Specificity analysis of SMO-KELM model.

presented model has offered a maximum sensitivity of 94.57% whereas the KNN, NB, SVM and DT models have led to minimal sensitivity of 92.60%, 87.90%, 83.20%, and 93.30%, respectively. Similarly, under the instance count of maximum 10,000, the presented model has obtained a higher sensitivity of 96.83%, whereas the KNN, NB, SVM, and DT models have led to minimal sensitivity of 93.60%, 89.10%, 84.20%, and 96%, respectively.

Figure 6.6 illustrates the analysis of the outcomes offered by the SMO-KELM method with respect to specificity. The figure shows that the SVM and NB methods have demonstrated ineffective classification show by achieving minimum specificity values. Also, the KNN method has obtained somewhat higher specificity value over the related techniques. Simultaneously, the DT method has exhibited competitive outcomes related to other techniques. However, the proposed SMO-KELM method has surpassed all the compared methods and attained highest specificity under each the instance. For instance, under the instance count of 2000, the proposed method has offered a highest specificity of 93.54% while the KNN, NB, SVM and DT approaches have led to least specificity of 84.20%, 83.40%, 80.20%, and 92.60%, respectively. Likewise, under the instance count of maximum 10,000, the projected method has attained a higher specificity of 92% but the KNN, NB, SVM and DT methods have led to minimum specificity of 89.30%, 86.40%, 84.30%, and 90.40%, respectively.

Figure 6.7 showcases the analysis of the outcomes offered by the SMO-KELM method with respect to accuracy. The figure stated that the SVM and NB methods have performed

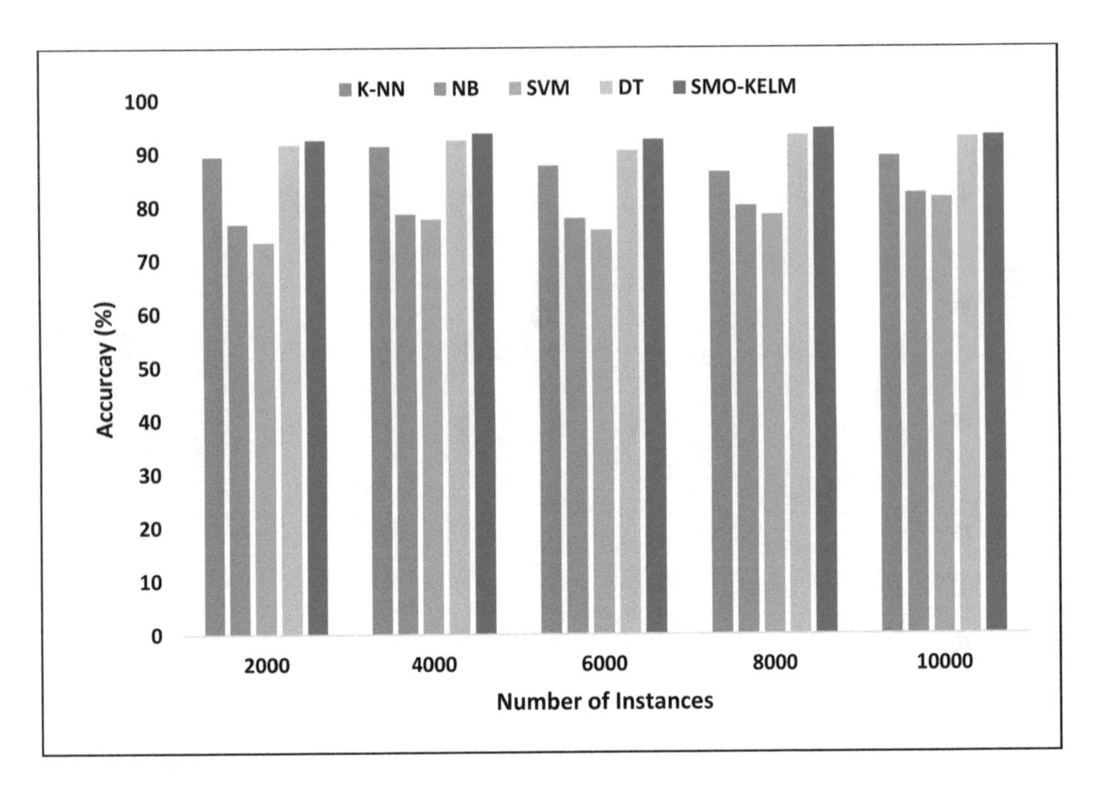

FIGURE 6.7 Accuracy analysis of SMO-KELM model.

ineffective classification performance by achieving minimum accuracy values. Also, the KNN method has attained somewhat higher accuracy value over the compared methods. Simultaneously, the DT approach has demonstrated competitive outcomes related to other techniques. However, the projected SMO-KELM method has surpassed each the related techniques and attained highest accuracy under all the instances. For instance, under the instance count of 2000, the proposed method has offered a highest accuracy of 92.45%, but the KNN, NB, SVM and DT models have led to least accuracy of 89.40%, 76.80%, 73.40%, and 91.60%, respectively.

Also, under the instance count of maximum 10,000, the proposed method has attained a higher accuracy of 93.18% while the KNN, NB, SVM and DT models have led to lest accuracy of 89.30%, 82.40%, 81.60%, and 92.80%, respectively.

6.4 CONCLUSION

This paper has developed an effective SMO-KELM model for disease diagnosis in IIoT environment. It incorporates data collection, data compression, disease diagnosis, and alert system. At the initial stage, the IoT devices attached to the patient's body are used to gather the patient data which is then compressed by the use of Deflate algorithm. Once the patient data has been compressed, it is transmitted to the cloud server via gateway devices in the fog layer. The cloud server performs decompression process and reconstructs the data proficiently. Then, the SMO-KELM model is executed to detect the presence of disease. Finally, an alarm will be raised in case of the occurrence of the diseases in real time to alert doctors, ambulance, and hospitals. The experimental validation of the SMO-KELM model is carried out and the results are observed under diverse aspects. In future, the performance of the SMO-KELM model can be further improvised by the use of feature selection methodologies.

REFERENCES

1. T. J. McCue, "$117 billion market for Internet of Things in healthcare by 2020," Forbes. Retrieved from https://www.forbes.com/sites/tjmccue/2015/04/22/117-billionmarket-for-internet-of-things-in-healthcare-by-2020%20Dec.%202015/#77b22a1971fa .
2. J. Mohammed, A. Thakral, A. F. Ocneanu, C. Jones, C.-H. Lung, and A. Adler, "Internet of Things: Remote patient monitoring using web services and cloud computing," In *Proc. IEEE Internet of Things (iThings)*, 1–3 Sept. 2014, Taipei, pp. 256–263.
3. S. M. Riazul Islam, D. Kwak, M. H. Kabir, M. Hossain, and K.-S. Kwak, "The Internet of Things for health care: A comprehensive survey," *IEEE Access*, vol. 3, pp. 678–708, 2015.
4. P. Y. Li, L. Guo, and Y. Guo, "Enabling health monitoring as a service in the cloud," In *Proc. IEEE Int. Conf. Conference on Utility and Cloud Computing*, Oct. 2014, pp. 81–84.
5. M. Hassanalieragh, A. Page, T. Soyata, G. Sharma, M. Aktas, G. Mateos, B. Kantarci, and S. Andreescu, "Health monitoring and management using Internet-of-Things (IoT) sensing with cloudbased processing: Opportunities and challenges," In *Proc. IEEE International Conference on Services Computing*, 2015, pp. 285–292.
6. J. Jara, M. A. Zamora-Izquierdo, and A. F. Skarmeta, "Interconnection framework for mHealth and remote monitoring based on the Internet of Things," *IEEE J. Sel. Areas Commun.*, vol. 31, no. 9, pp. 47–65, Sep. 2013.
7. B. Xu, L. D. Xu, H. Cai, C. Xie, J. Hu, and F. Bu, "Ubiquitous data accessing method in IoT-based information system for emergency medical services," *IEEE Trans. Industr. Inform.*, vol. 10, no. 2, May 2014.
8. K. Zhang, K. Yang, X. Liang, Z. Su, X. (Sherman) Shen, and H. H. Luo, "Security and privacy for mobile healthcare networks: From a quality of protection perspective," *IEEE Wireless Commun. Mag.*, vol. 22, pp. 104–112, Aug. 2015.
9. J. Granados, A.-M. Rahmani, P. Nikander, P. Liljeberg, and H. Tenhunen, "Web-Enabled Intelligent Gateways for eHealth Internet-of-Things," In *International Internet of Things Summit*, pp. 248–254, Springer, Cham.
10. S. Amendola, R. Lodato, S. Manzari, C. Occhiuzzi, and G. Marrocco, "RFID technology for IoT-based personal healthcare in smart spaces," *IEEE Internet Things J.*, vol. 1, no. 2, April 2014.
11. L. Catarinucci, D. De Donno, L. Mainetti, L. Palano, L. Patrono, M. L. Stefanizzi, and L. Tarricone, "An IoT-aware architecture for smart healthcare systems," *IEEE Internet Things J.*, 2015. DOI 10.1109/JIOT.2015.2417684.
12. A. Sawand, S. Djahel, Z. Zhang, and F. Näıt-Abdesselam, "Toward energy-efficient and trustworthy eHealth monitoring system," *China Commun.*, vol. 12, pp. 46–65, January 2015.
13. M. Chen, Y. Zhang, Y. Li, S. Mao, and V. Leung, "EMC: Emotionaware mobile cloud computing in 5G," *IEEE Network*, vol. 29, no. 2, pp. 32–38, 2015.
14. M. Chen, Y. Zhang, Y. Li, M. Hassan, and A. Alamri, "AIWAC: Affective interaction through wearable computing and cloud technology," *IEEE Wirel. Commun. Mag.*, vol. 22, no. 1, pp. 20–27, 2015.
15. L. Hu, M. Qiu, J. Song, and M. Shamim Hossain, "Software defined healthcare networks," *IEEE Wirel. Commun. Mag.*, vol. 22, no. 6, Dec. 2015.

16. Y. Kim, S. Choi, J. Jeong, and Y. H. Song, "Data dependency reduction for high-performance FPGA implementation of DEFLATE compression algorithm," *J. Syst. Archit.*, vol. 98, pp. 41–52, 2019.
17. Q. Li, H. Chen, H. Huang, X. Zhao, Z. Cai, C. Tong, W. Liu, and X. Tian, "An enhanced grey wolf optimization based feature selection wrapped kernel extreme learning machine for medical diagnosis," *Comput. Math. Methods Med.*, vol. 2017, 2017.
18. N. Khare, P. Devan, C. L. Chowdhary, S. Bhattacharya, G. Singh, S. Singh, and B. Yoon, "SMO-DNN: Spider monkey optimization and deep neural network hybrid classifier model for intrusion detection," *Electronics*, vol. 9, no. 4, p. 692, 2020.

CHAPTER 7

Fuzzy Support Vector Machine with SMOTE for Handling Class Imbalanced Data in IoT-Based Cloud Environment

A. Francis Saviour Devaraj,[1] P. Vijayakarthik,[2] S. Dhanasekaran,[1] G. Murugaboopathi,[1] and B.S. Murugan[1]

[1]*Department of Computer Science and Engineering, School of Computing, Kalasalingam Academy of Research and Education, Krishnankoil, Tamil Nadu, India*

[2]*Department of Information Science and Engg, Sir. M. Visvesvaraya Institute of Technology, Bangalore, India*

CONTENTS

7.1 INTRODUCTION

Internet of Things (IoT) is assumed as an interlinked network of modern sensors that has minimum storage as well as computing ability. IoT in conjunction with cloud computing (CC) has enormous benefits like memory and adequate computing energy, which facilitates the required services such as medical sector [1, 2] and smart cities. Hence, observation as well as interacting with patients remotely is highly essential in this approach. In addition, the requirement to offer minimum cost, maximum superiority, and patient-based smart healthcare for the individuals is developed. The evolution of IoT [3] and CC models [4] are required in real-time, modern, and remote medical services for smart cities. Moreover, the combination of IoT and CC methods offers unique and tremendous benefits in smart healthcare monitoring. Recently, humans in smart cities have permission to apply modern sensor devices as well as latest mobile techniques. In a smart city, identifying a medical expert, hospitals, and pharmacies are highly a dark room. Also, a patient suffering from severe disease cannot move quickly to hospitals.

In order to resolve these issues, a smart healthcare-monitoring approach has been developed by combining the available resources which enhance the superiority and accessibility of medical services. Using the smart healthcare-monitoring model, the healthcare-oriented multimedia signals are transmitted from modern sensing devices and mobile tools to offer periodic guidance and qualified medical services to users. These medical data as well as signals are massive in size and complex to manage due to the complexity. The clinical sector has been introduced with massive requirements for industrial sector.

Even though IoT provides immediate and better treatments, it also meets maximum economic reviews for government as well as private sector. Recently, modern clinical centers have found a competition between diverse medical providers in offering better and comfortable facilities with good accuracy, dependability, and minimum expense [5]. Thus, the combination of CC and IoT in healthcare concentrates in better study and services. Various models of IoT tools have been developed in the field of healthcare such as portable tools like BP devices, portable insulin syringe, stress tracking machine, weight observing and fitness tools, hearing devices, and EEG and ECG monitors.

Though it has massive advantages, mobile approaches and ICT have resulted in tremendous benefits that are carried out in medical filed. Remote patient-monitoring function is performed using wireless and ubiquitous sensing model. A practical health-monitoring framework termed as Healthcare Industrial IoT has been projected in Ref. [6]. The newly deployed method holds vital capabilities for examining patient's medical information and lowers the mortality rates. Also, it gathers the patient details under the application of diverse healthcare tools and sensors. Followed by, privacy is considered as a major issue in the transmission of patient's medical data to CC by doctors. Moreover, the theft action or clinical flaws can be identified by physicians using the strategies such as signal enhancement and watermarking methodologies. Gope and Hwang [7] applied the properties of IoT method to body sensor network (BSN). Here, patient will be observed with the help of tiny-powdered as well as light-weight sensor networks. Also, it serves an effective security and saves the patient details; a protective IoT-based medical network is described named as BSN-Care.

A practical health-monitoring approach is presented in Ref. [8] for distant heart patients under the variables. It is designed for offering the interface among a physician and patients and it facilitates two-way interactions. In addition, the application of BSN can be improved by providing warning alerts for serious cases, which are transmitted to remote users. An IoT-based, noninvasive monitoring system is presented in Ref. [9] for urban medical system. The structure of u-healthcare is composed of BSN; modern medical server and hospital system are considered to be vital units that describe the model. Hussain et al. [10] utilized a patient-based sensing method for aged and handicapped persons. The key purpose of this approach is to offer the service-centric response in serious condition of a user. Kim and Chung (2015) presented a serious case-monitoring technology with the help of context motion observation for patients who suffer from chronic diseases. It examines the recent condition of a patient on the basis of contextual data and offers essential data by investigating the patient's lifestyle.

Catarinucci et al. [11] introduced latest techniques in IoT-based medical applications. It is surveyed that the existing platform, application, and trends in IoT-based medical solution are significant. Xu et al. [12] managed the heterogeneity issue of data type in IoT environment under the application of data model. Furthermore, resource-driven data-accessing technology has been developed for gaining and computing IoT data exclusively for the purpose of enhancing permission of IoT resources. Also, IoT-based system is applicable in handling critical situations and has been established to show the way of collection, combination, and interoperate IoT data elasticity. Box et al. [13] deployed and executed smart-home-based environment like iHome Health-IoT. It is embedded with open-platform relied intelligent clinical and improved connectivity for integration of tools and services. In addition, modern pharmaceutical packages as well as biomedical sensor-based tools are embedded in presented approach. Maia et al. [14] depicted an EcoHealth, a web middleware model for correlating persons with experts with the help of portable body sensors. It combines data attained from heterogeneous sensors. Then, the concatenated data is applied to send the alert message about patient's state and significant symbols practically under the application of Internet service.

A viable solution is to apply the health conditions and find the name of a disease from health data gathered with the help of personal IoT devices [15]. A prototype has been developed for diagnosing disease by converting the disease diagnosing models into machine readable format. Hajihashemi et al. [16] introduced a novel model to process the affinity of two multiattribute time series on the basis of temporal model of Smith–Waterman, named as bioinformatics model. It mimics the complexities based on data irregularity and aggregation, which emerge while processing the sensed data. This model can be validated by using integrated data collected from electronic health records (EHR) and non-wearable's fixed at apartments in the United States. The accomplished outcome implies that different health patterns have been examined to predict the anomalies in humans. Liu et al. [17] proposed a new motif discovery model for large-scale time series, named as MDLats. Motifs are identical sequences that are important in examining ECG patterns of cardiac patients. Hence, it reapplies the previous data to a greater extent and makes use of a relation among traditional data as well as current information.

This study devises an efficient fuzzy support vector machine (FSVM) with Synthetic Marginal Oversampling Technique (SMOTE) model called SMOTE-FSVM for class imbalance problem in IoT- and cloud-based disease diagnosis. The proposed SMOTE-FSVM model initially involves data collection, upsampling, and data classification. At the first stage, the data collection process takes place using IoT devices connected to patients and the data is transmitted to cloud server. Then, in the second stage, SMOTE-based upsampling of marginal data samples takes place by the generation of synthetic data. Finally, in the third stage, the medical data classification process is carried out using FSVM model. The proposed SMOTE-FSVM model has been assessed using PIMA Indians Diabetes dataset and the experimental validations are investigated under distinct performance measures.

7.2 THE PROPOSED MODEL

The working operation of the SMOTE-FSVM model is depicted in Figure 7.1. As shown, the medical data has been gathered from a diverse set of sources, namely IoT sensor data, medical records, and University of California, Irvine (UCI) repository. The patient's data are acquired and transmitted to the cloud. Then, SMOTE-based upsampling process takes place to resolve the class imbalance problem. Finally, FSVM-based data classification task is carried out to determine the class labels and find the existence of the diseases.

7.2.1 SMOTE Model

SMOTE method has been established for neutralizing the irregular dataset issue for classification [18]. It synthesizes the instances of minimal class under the operation in "feature space" instead of "data space." It is considered as upsampling method where the marginal data produces N% of synthetic data. The ratio enhances marginal data which is comparable with majority data. It enhances the samples of marginal data and extends the decision causes for classifiers.

Here, few variables like T, N%, and k are invoked initially, where T defines the value of marginal class instances, N% indicated the ratio of upsampling, and k refers to the k value of k nearest neighbor (kNN) of specific marginal class instances. The procedure contributes to produce synthetic instances as given in the following points:

- Once the initialization is completed, a marginal class instance has been selected where synthetic data should be produced.
- After this one among kNN marginal class neighbors of instances can be selected randomly.

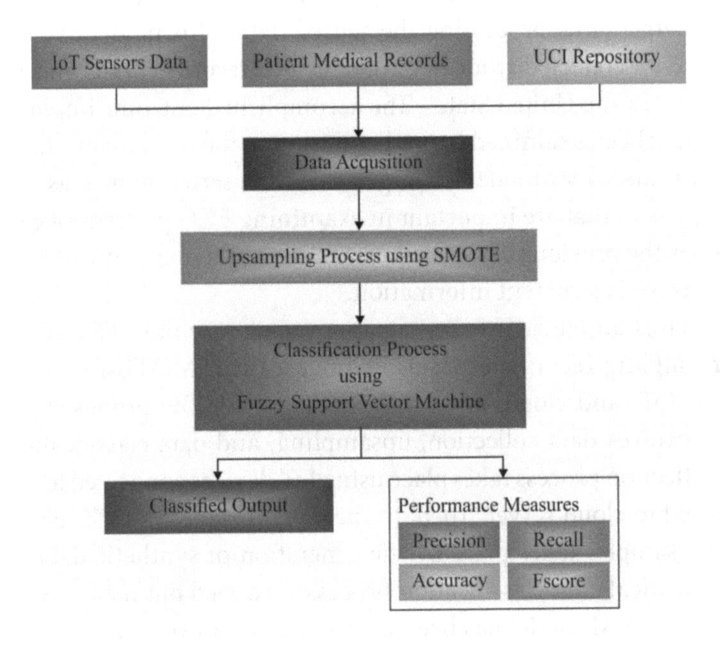

FIGURE 7.1 Framework of SMOTE-FSVM model.

- It is known that an instance is composed of massive feature data, so a synthetic instance is generated by producing synthetic data for all feature data. A synthetic data is produced by including a factor to primary feature data. It is estimated in two steps. Initially, the chosen feature data is reduced from first marginal feature data. Then, the reduced value is improved with a value from 0 and 1.
- It is computed for feature data of a specific marginal class instance that produces a row of synthetic instance for marginal class instance.
- For N% upsampling, it is computed for rounded value of (N/100) to adjacent integer. It produces N% of upsampling of individual marginal class instance.
- This strategy is performed for T marginal class instances and provides N% upsampling of all marginal class instances.

7.2.2 FSVM-Based Classification Model

Generally, the SVM model is applied for classification issues. Assume that there are set of labeled training points:

$$(y_1, x_1), \ldots, (y_l, x_l) \tag{7.1}$$

Every training point $x_i \in R^N$ comes under two classes and which is provided as a label $y_i \in \{-1,1\}$ for $i = 1,\ldots,l$. Mostly, exploring an applicable hyperplane for input space is highly limited practically. A better solution for this problem is mapping an input space to high-dimension feature space as well as finding a best hyperplane. Suppose $z = \varphi(x)$ represents the feature space vector along with a mapping φ from R^N to feature space $\mathcal{Z}$. The hyperplane can be determined as:

$$w \cdot z + b = 0 \tag{7.2}$$

Described by a pair (w, b), where it isolates the point x_i on the basis of given function,

$$f(x_i) = sign(w \cdot z_i + b) = \begin{cases} 1, & if \ y_i = 1 \\ -1, & if \ y_i = -1 \end{cases} \tag{7.3}$$

where $w \in \mathcal{Z}$ and $b \in R$. In depth, the set S is meant to be linearly separable when it has (w,b) where the uncertainties are valuable for elements of set S.

$$\begin{cases} (w \cdot z_i + b) \geq 1, & if \ y_i = 1 \\ (w \cdot z_i + b) \leq -1, & if \ y_i = -1, \end{cases} \quad i = 1,\ldots,l \tag{7.4}$$

For linearly separable set S, it identifies an exclusive and best hyperplane and the margin among the projections of training points of two different classes are improved. When the set S is not linearly separable, classification violations should be activated in SVM formulation. In order to overcome these not linearly separable problems, the existing analysis could be normalized by establishing nonnegative variables $\xi_i \geq 0$, where equation (7.4) is changed to:

$$y_i\left(w \cdot z_i + b\right) \geq 1 - \xi_i, i = 1,\ldots,l \tag{7.5}$$

The nonzero ξ_i in equation (7.5) for the point x_i would not satisfy equation (7.4). Hence, the term $\Sigma_{i=1}^{l}\xi_i$ can be referred to as a measure of misclassifications. The best hyperplane issue is regarded as solution to a problem.

$$\begin{aligned} &minimize \frac{1}{2} w \cdot w + C\sum_{i=1}^{l}\xi_i \\ &subject\ to\ y_i(w \cdot z_i + b) \geq 1 - \xi_i, i = 1,\ldots,l \\ &\xi_i \geq 0, i = 1,\ldots,l \end{aligned} \tag{7.6}$$

where C implies a constant. The parameter C is assumed to be a regularization parameter. It is a free parameter in SVM formulation. Changing this parameter develops a balance among margin maximization as well as classification violation. Detail definitions are identified in Refs. [4, 6]. Exploring best hyperplane in equation (7.6) is said to be QP problem that is resolved by developing Lagrangian and changed into the dual.

$$\begin{aligned} &maximize\ W(\alpha) = \sum_{i=1}^{l}\alpha_i - \frac{1}{2}\sum_{i=1}^{l}\sum_{j=1}^{l}\alpha_i\alpha_j y_i y_j z_i \cdot z_j \\ &subject\ to \sum_{i=1}^{l} y_i\alpha_i = 0\ 0 \leq \alpha_i \leq C, i = 1,\ldots,l \end{aligned} \tag{7.7}$$

where $\alpha = (\alpha_1,\ldots,\alpha_l)$ denotes a vector of nonnegative Lagrange multipliers linked with the constraints in equation (7.5). The Kuhn–Tucker statement is one of the well-known models in SVM. Based on this method, a solution $\bar{\alpha}_i$ of problem presented in equation (7.7) meets

$$\bar{\alpha}_i\left(y_i\left(\bar{w} \cdot z_i + \bar{b}\right) - 1 + \bar{\xi}_i\right) = 0, i = 1,\ldots,l \tag{7.8}$$

$$\left(C - \bar{\alpha}_i\right)\bar{\xi}_i = 0, i = 1,\ldots,l \tag{7.9}$$

From the equality, the nonzero values $\bar{\alpha}_i$ in equation (7.8) are named as constraints in equation (7.5) and are computed with the equality sign. The point x_i is corresponding with $\bar{\alpha}_i > 0$ and is termed as support vector. However, it has two types of support vectors in a nonseparable case. For $0 < \bar{\alpha}_i < C$, the corresponding support vector x_i meets the equalities $y_i\left(\bar{w} \cdot z_i + \bar{b}\right) = 1$ and $\xi_i = 0$. For, $\bar{\alpha}_i = C$, the corresponding ξ_i is nonempty for corresponding support vector x_i which does not meet equation (7.4). It refers the support vectors as errors. The point x_i corresponds with $\bar{\alpha}_i = 0$ is categorized accurately and clearly for decision margin. In order to develop an optimal hyperplane $\bar{w} \cdot z_i + \bar{b}$, the following notion has been applied:

$$\bar{w} = \sum_{i=1}^{l} \bar{\alpha}_i y_i z_i \tag{7.10}$$

and a scalar $\bar{b}$ is computed from the Kuhn–Tucker conditions in equation (7.8). The decision function is generalized from equation (7.3) as well as equation (7.10) so that,

$$f(x) = sign(w \cdot z + b) = sign\left(\sum_{i=1}^{l} \alpha_i y_i z_i \cdot z + b\right) \tag{7.11}$$

As there is no existence of prior knowledge of φ, the processing issue in equations (7.7) and (7.11) is not possible. Also, an optimal feature of SVM which is not essential to learn regarding φ. It is essential for a function $K(\cdot,\cdot)$ named as kernel which process the dot product of data points in feature space $\mathcal{Z}$, which is:

$$z_i \cdot z_j = \varphi(x_i) \cdot \varphi(x_j) = K(x_i, x_j). \tag{7.12}$$

The performance which meets the Mercer's statement could be applied as dot products and applied as kernels. The application of polynomial kernel of degree d

$$K(x_i, x_j) = (1 + x_i \cdot x_j)^d \tag{7.13}$$

Hence, the nonlinear separating hyperplane is referred as a solution of,

$$maximize\ W(\alpha) = \sum_{i=1}^{l} \alpha_i - \frac{1}{2} \sum_{i=1}^{l} \sum_{j=1}^{l} \alpha_i \alpha_j y_i y_j K(x_i, x_j)$$

$$subject\ to\ \sum_{i=1}^{l} y_i \alpha_i = 0\ 0 \le \alpha_i \le C, i = 1, \ldots, l \tag{7.14}$$

and a decision function is provided as:

$$f(x) = sign(w \cdot z + b) = sign\left(\sum_{i=1}^{l} \alpha_i y_i K(x_i, x) + b\right) \tag{7.15}$$

In classical SVM, every data point is assumed with identical significance and allocated the similar penal variable in objective function. Hence, in real-time classification domains, few sample points, like noises, cannot be allocated to any class, and every instance does not contain equal meaning to decision surface. In order to resolve these issues, the principle of FSVM was developed in Refs. [19]. Fuzzy membership to all sample points is established in which various sample points make diverse contributions for developing the decision surface. Let, the training samples are:

$$S=\{(x_i,y_i,s_i),\ i=1,\ldots,N\}, \tag{7.16}$$

where $x_i \in R^n$ denotes the n-dimension sample point, $y_i \in \{-1,+1\}$ refers the class labels, and s_i ($i = 1,\ldots, N$) implies a fuzzy membership that meets $\sigma \leq s_i \leq 1$ using adequately small constant $\sigma > 0$. The quadratic optimization (QP) issue for classification is assumed in the following:

$$\min_{w,s,\xi} \frac{1}{2} w^T w + C \sum_{i-1}^{l} s_i \xi_i \tag{7.17}$$

$$s.t.\ y_i(w^T x_i + b) \geq 1 - \xi_i,\ \xi_i \geq 0,\ i=1,\ldots,l,$$

where w defines normal vector of isolating hyperplane, b implies a bias term, and C shows a parameter that has to be calculated in advance for controlling the tradeoff among a classification margin as well as expense of misclassification error. As s_i is an attitude of parallel point x_i to one class and a slack variables ξ_i are value of error, and a term $s_i\xi_i$ is assumed as measure of error with various weights. It is pointed that bigger s_i is more significant for corresponding point; the smaller the s_i, minimum is the significance for corresponding point; hence, diverse input points make various contributions to learn the decision surface. However, FSVM finds maximum and robust hyperplane by increasing the margin by reducing the misclassification error.

For resolving the FSM optimal issue, equation (7.17) is converted into a dual problem under the application of Lagrangian multipliers α_i:

$$\max_{\alpha} \sum_{i=1}^{N} \alpha_i - \frac{1}{2} \sum_{i=1}^{N} \sum_{j=1}^{N} \alpha_i \alpha_j y_i y_j x_i x_j \tag{7.18}$$

$$s.t. \sum_{i=1}^{N} y_i \alpha_i = 0,\ 0 \leq \alpha_i \leq s_i C,\ i=1,\ldots,N,$$

When compared to standard SVM, the predefined statement contains minimum variations that are an upper bound of measures of α_i. By resolving the dual problem in equation (7.18) for optimal α_i, w and b could be eliminated as same as standard SVM.

7.3 SIMULATION RESULTS AND DISCUSSION

This section examines the classification performance obtained by the SMOTE-FSVM model using PIMA Indians Diabetes Dataset. It comprises a total of 768 instances with 8 attributes and 2 classes. A set of 34.90% of samples comes under positive class and the remaining 65.10% of samples come under negative class. The details are listed out in Table 7.1. Among the total number of 500 negative instances and 268 positive instances, the SMOTE model upsamples the data instances and convert 536 instances as positive. Thereby, the number of instances in the two class labels is effectively balanced.

Figure 7.2 analyzes the performance of the SMOTE-FSVM model in terms of precision. The figure portrayed that the DT model is found to be the ineffective performer which has attained a minimal precision of 81.40%. Next to that, the Logitboost model has outperformed the DT model by attaining a precision of 84.60%. Besides, the SVM model has reached to a certainly higher precision of 85.32%. In line with, the LR and FSVM models have led to near acceptable precision values of 88% and 87.46%, respectively. However, the presented SMOTE-FSVM model has achieved better performance by attaining maximum precision of 91.47%.

Figure 7.3 analyzes the performance of the SMOTE-FSVM model in terms of recall. The figure portrayed that the LogitBoost model is found to be the ineffective performer which has attained a minimal recall of 77.61%. Next to that, the DT model has outperformed the earlier model by attaining a recall of 79.02%. Besides, the LR model has reached to a certainly higher recall of 79.27%. In line with, the SVM and FSVM models have led to near acceptable recall values of 80.12 and 82.99%, respectively. However, the presented SMOTE-FSVM model has achieved better performance by attaining maximum recall of 89.63%.

Figure 7.4 analyzes the performance of the SMOTE-FSVM model in terms of accuracy. The figure portrayed that the DT model is found to be the ineffective performer which has attained a minimal accuracy of 73.82%. Next to that, the LogitBoost model has outperformed the DT model by attaining an accuracy of 74.08%. Besides, the LR model has reached to a certainly higher accuracy of 77.21%. In line with, the SVM and FSVM models have led to near acceptable accuracy values of 76.65% and 80.11%, respectively. However, the presented SMOTE-FSVM model has achieved better performance by attaining maximum accuracy of 85.52%.

Figure 7.5 analyzes the performance of the SMOTE-FSVM model in terms of F score. The figure portrayed that the DT model is found to be the ineffective performer which has attained a minimal F score of 80.19%. Next to that, the Logitboost model has outperformed

TABLE 7.1 Dataset Description

Description	Value
Number of samples	768
Number of features	8
Number of classes	2
% of positive samples	34.90%
% of negative samples	65.10%
Data source	[20]

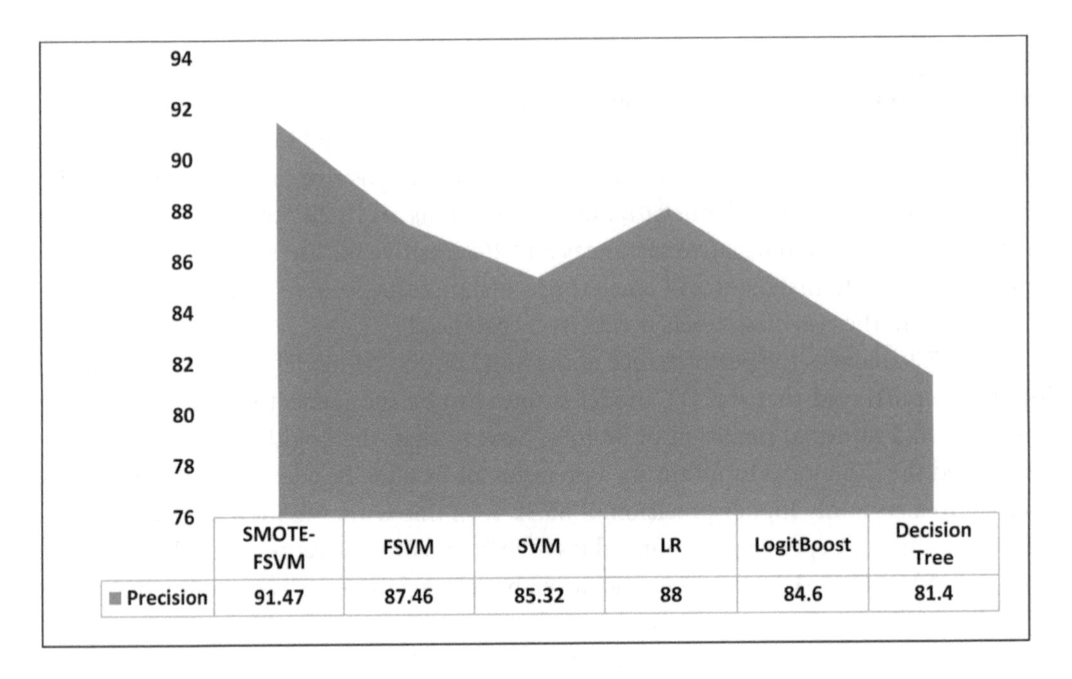

FIGURE 7.2 Precision analysis of SMOTE-FSVM model.

the DT model by attaining an F score of 80.95%. Besides, the LR model has reached to a certainly higher F score of 83.41%. In line with, the SVM and FSVM models have led to near acceptable F score values of 85.83% and 84.57%, respectively. However, the presented SMOTE-FSVM model has achieved better performance by attaining maximum F score of 90.45%.

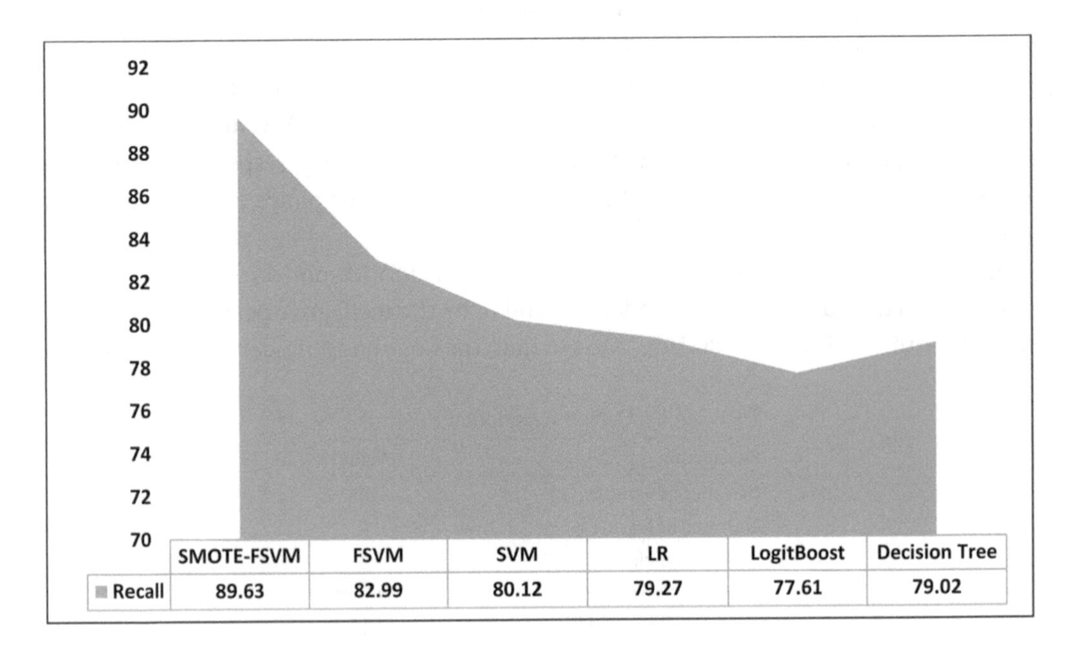

FIGURE 7.3 Recall analysis of SMOTE-FSVM model.

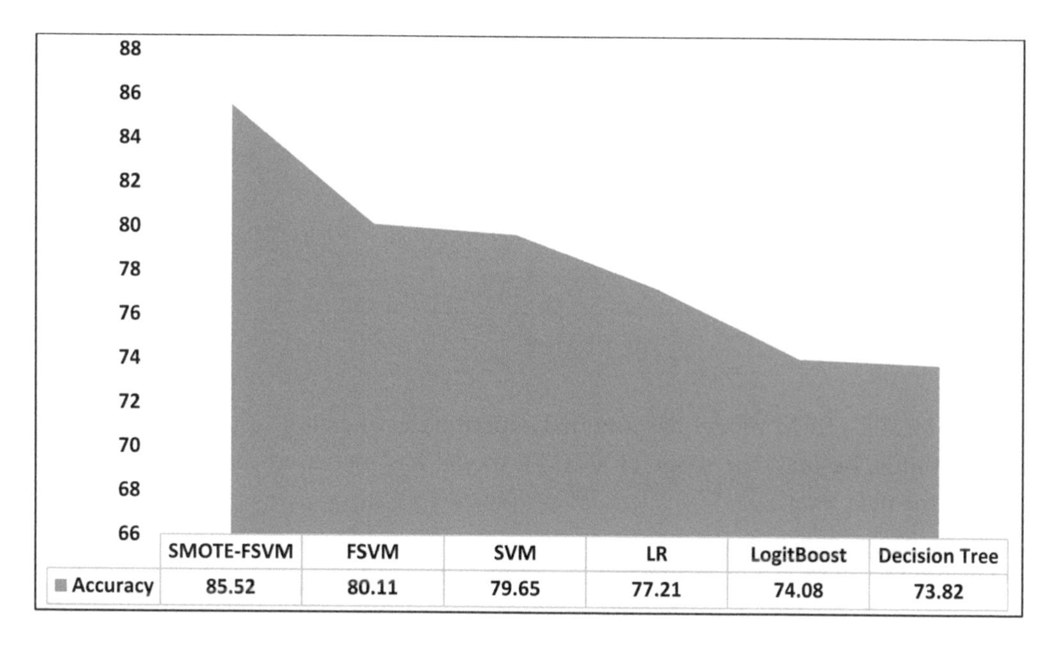

FIGURE 7.4 Accuracy analysis of SMOTE-FSVM model.

Table 7.2 provides the response time analysis of SMOTE-FSVM model over the compared models. The table values denoted that the LogitBoost model has required high computation time of 40 seconds and offered higher response time compared to other models. Besides, the LR and DT models have attained somewhat lower response times of 36 and 35 seconds, respectively. Followed by, the proposed SMOTE-FSVM model has achieved moderate response time of 32 seconds which is higher than previous method not than FSVM and SVM models. However, the FSVM and SVM models have required minimal response time of 28 and 25 seconds, respectively. The experimental validation indicated

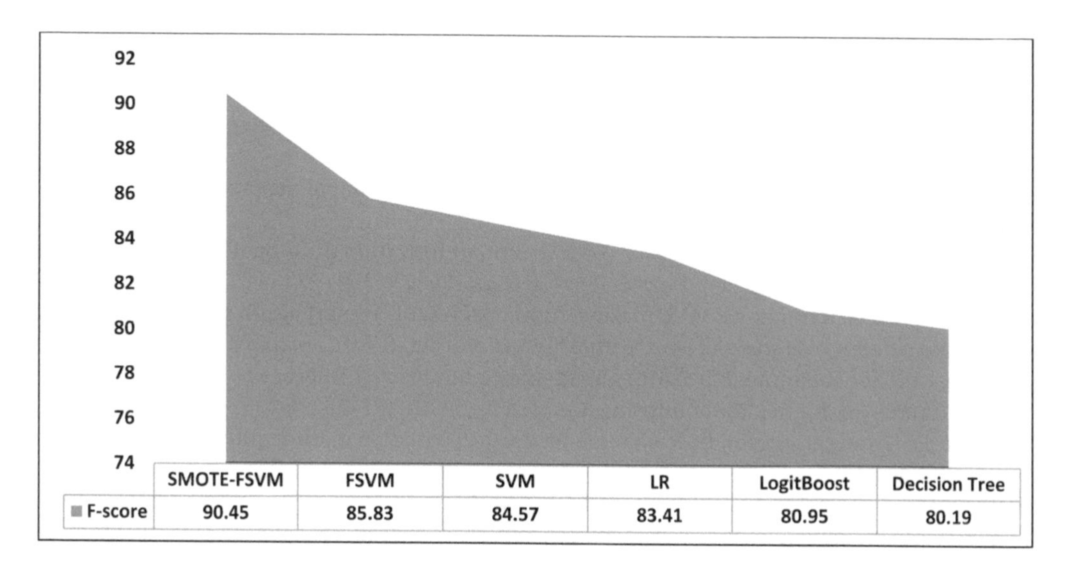

FIGURE 7.5 F score analysis of SMOTE-FSVM model.

TABLE 7.2 Response Time Analysis

Classifiers	Time (second)
SMOTE-FSVM	32
Fuzzy support vector machine	28
Support vector machine	25
Logistic regression	36
LogitBoost	40
Decision tree	35

that the SMOTE-FSVM model has attained superior classification results over the compared methods. Besides, the usage of SMOTE model has increased the classifier performance to the next level.

7.4 CONCLUSION

This chapter has developed a novel SMOTE-FSVM model for class imbalance problem in IoT- and cloud-based disease diagnosis. The medical data has been gathered from a diverse set of sources, namely IoT sensor data, medical records, and UCI repository. The patient's data are acquired and transmitted to the cloud. Then, SMOTE-based upsampling process takes place to resolve the class imbalance problem. Finally, FSVM-based data classification task is carried out to determine the class labels and find the existence of the diseases. The classification performance of the SMOTE-FSVM model has been validated using PIMA Indians Diabetes dataset. The simulation outcome verified that the SMOTE-FSVM model has obtained maximum results with the precision of 91.47%, recall of 89.63%, accuracy of 85.52%, and F score of 90.45%.

REFERENCES

1. Yin Y, Zeng Y, Chen X, Fan Y (2016) The internet of things in healthcare: an overview. *J Ind Inf Integr* 1:3–13.
2. Chen M, Zhang Y, Qiu M, Guizani N, Hao Y (2018) SPHA: smart personal health advisor based on deep analytics. *IEEE Commun* 56(3):164–169.
3. Hu L et al. (2016) Internet of Things cloud: architecture and implementation. *IEEE Commun Mag* 54(12 Supp):32–39.
4. Chen M et al. (2018) Edge-CoCaCo: Toward joint optimization of computation, caching, and communication on edge cloud. *IEEE Wireless Commun* 25(3):21–27.
5. Muhammad G, Rahman SKMM, Alelaiwi A, Alamri A (2017) Smart health solution integrating IoT and cloud: a case study of voice pathology monitoring. *IEEE Commun Mag* 55(1):69–73.
6. Hossain MS, Muhammad G (2016) Cloud-assisted industrial Internet of Things (IIoT) enabled framework for health monitoring. *Comput Netw* 101:192–202.
7. Gope P, Hwang T (2016) BSN-Care: a secure IoT-based modern healthcare. *IEEE Sens J* 16(5):1368–1376.
8. Priyanka K, Tripathi NK (2015) A real-time health monitoring system for remote cardiac patients using smartphone and wearable sensors. *Int J Telemed Appl.* doi: 10.1155/2015/373474
9. Gelogo YE, Hwang HJ, Kim H (2015) Internet of Things (IoT) framework for u-healthcare system. *Int J Smart Home* 9(11):323–330.

10. Hussain A, Wenbi R, Lopes A, Nadher M, Mudhish M (2015) Health and emergency-care platform for the elderly and disabled people in the Smart City. *J Syst Softw* 110:253–263.
11. Catarinucci L, De Donno D, Mainetti L, Palano L, Patrono L, Stefanizzi ML, Tarricone L (2015) An IoT-aware architecture for smart healthcare systems. *IEEE Internet Things* 2(6):515–526.
12. Xu B, Da Xu L, Member S, Cai H, Xie C, Hu J, Bu F (2014) Ubiquitous data accessing method in IoT-based information system for emergency medical services. *IEEE Trans Ind Inf* 10(2):1578–1586.
13. Box IM, Yang G, Xie L, Mäntysalo M, Zhou X, Pang Z, Da Xu L, Member S (2014) A health-IoT platform based on the integration of intelligent packaging, unobtrusive bio-sensor and intelligent medicine box. *IEEE Trans Ind Inf* 10(4):2180–2191.
14. Maia P, Batista T, Cavalcante E, Baffa A, Delicato FC, Pires PF, Zomaya A (2014) A web platform for interconnecting body sensors and improving health care. *Proced Comput Sci* 40:135–142
15. La HJ (2016) A conceptual framework for trajectory-based medical analytics with IoT contexts. *J Comput Syst Sci* 82(4):610–626.
16. Hajihashemi Z, Popescu M, Member S (2016) A multidimensional time-series similarity measure with applications to eldercare monitoring. *IEEE J Biomed Health Inf* 20(3):953–962.
17. Liu B, Li J, Chen C, Tan W, Member S, Chen Q, Zhou M (2015) Efficient motif discovery for large-scale time series in healthcare. *IEEE Trans Ind Inf* 11(3):583–590.
18. Chawla NV, Bowyer KW, Hall LO, Kegelmeyer WP (2002) SMOTE: synthetic marginal over-sampling technique. *J Artif Int Res* 16:321e357.
19. Lin CF, Wang SD. (2002) Fuzzy support vector machines. *IEEE Trans Neural Netw* 13(2):464–471.
20. https://www.kaggle.com/uciml/pima-indians-diabetes-database.

CHAPTER 8

Energy-Efficient Unequal Clustering Algorithm Using Hybridization of Social Spider with Krill Herd in IoT-Assisted Wireless Sensor Networks

R. Buvanesvari[1] and A. Rijuvana Begum[2]

[1]*Department of Information Technology, PRIST Deemed University, Thanjavur, Tamil Nadu, India*

[2]*Department of Electronics and Communication Engineering, PRIST Deemed University, Thanjavur, Tamil Nadu, India*

CONTENTS

8.1 INTRODUCTION

At present, the Internet of Things (IoT) is the interconnection of exclusively recognizable embedded computing devices in the available Internet framework. Wireless sensor networks (WSNs) comprise numerous sensor nodes which undergo deployment in various applications such as medicine, smart industry, surveillance, precision agriculture, etc. [1, 2]. Besides, WSN offered an essential part in the emergence of technologies, e.g., big data, cloud, and IoT [3]. On the other hand, these have several design limitations because of the high processing and energy limitations. As a result, energy-efficient clustering techniques have emerged. Normally, WSN comprises hundreds to thousands of sensor nodes and base station (BS). The nodes in WSN transmit data to BS independently or might create diverse clusters with cluster heads (CHs) [4]. The data transmission between the nodes and BS takes place in a direct and indirect way [5]. The architecture of clustering is shown in Figure 8.1. The issue in single-hop data transmission is the excessive energy utilization because of the massive transmission range [6]. At the same time, the nodes which are present nearer to the BS expire rapidly in multi-hop data transmission as they transmit every packet of the network to BS. It is called as hot spot issue and several techniques have been presented for mitigating it by the creation of unequal clusters with respect to size [7].

Clustering as well as unequal clustering models has been employed to achieve energy efficiency in WSN. The clustering process in WSN takes place by merging the sensor nodes to clusters set. The clustering process enables the CHs to gather the data from the cluster members (CM) [5, 8]. The utilization of energy by the CHs closer to the BS is higher than the CHs far from the BS. It implies that the CHs closer to BS are far from the CHs located away from the BS due to the presence of intra-cluster data transmission from its CM, data aggregation, and inter-cluster data from other CHs to relay data to BS. It affects the network connectivity, and the clusters closer to BS lead to the coverage issue which is known as hot spot issue.

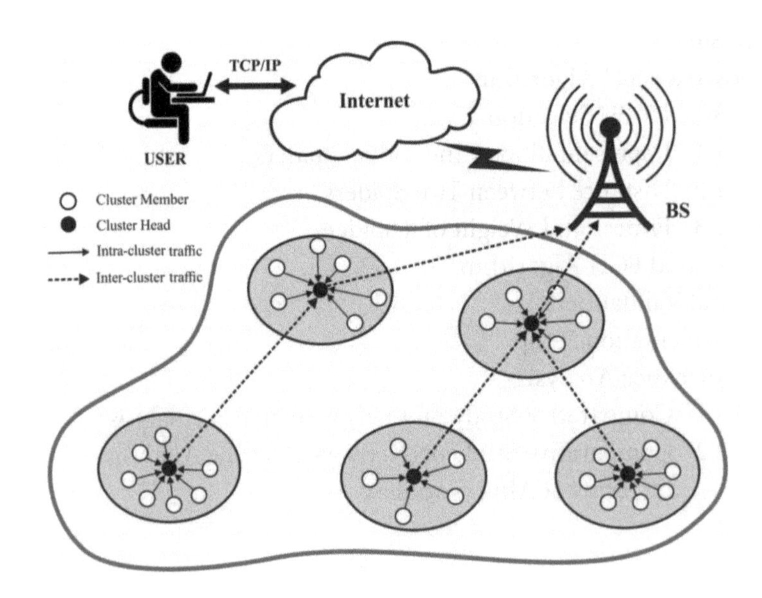

FIGURE 8.1 Clustering process in wireless sensor network.

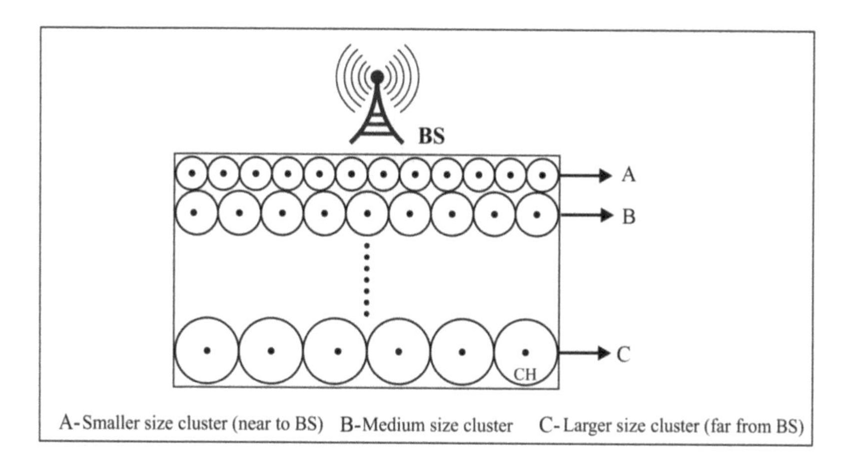

FIGURE 8.2 Unequal clustering process.

The unequal clustering approach is an effective way that deals with the hot spot issue due to the fact that it can be employed to balance the load between the CHs. The aim of unequal clustering technique is identical to the equal clustering one with extra feature of achieving energy efficiency and resolving hot spot issue [9]. The unequal clustering organizes the cluster size based on its distance to BS. The architecture of unequal clustering is shown in Figure 8.2. A cluster with least size implies lower number of CM and lesser intra-cluster data transmission [10]. So, smaller-sized clusters can spend the energy for inter-cluster data transmission and CHs cannot exhaust their energy quickly. In case of long distance to BS, the cluster size will be increased. When the cluster has more number of CMs, large amount of energy will be utilized for intra-cluster data transmission [11]. Since the cluster is located far from the BS, inter-cluster data transmission will be low and there is no requirement to utilize high energy for routing data between clusters.

The unequal clustering model will force every CH to utilize an identical amount of energy; so, the CHs near or far from the BS utilize identical quantity of energy. In addition, the construction of clusters might create a two-level hierarchy with high as well as low levels. The sensing devices will send the data regularly to its respective CH which will perform data aggregation and send it to the BS in a straightforward way or intermittent CHs. In case of a death of a CH or if it moves to other clusters, the reclustering process will take place among the nodes to elect new CHs.

This chapter introduces a new hybridization of Social Spider (SS) with Krill Herd (KH) algorithm named SS-KH for unequal clustering in WSN. In this case, the SS algorithm firstly selects the tentative cluster heads (TCH) and then the KH algorithm is applied to decide the final cluster heads (FCH). The presented SS-KH algorithm successfully selects the CHs in a proficient way. This algorithm undergoes diverse scenarios based on the positions of CHs. Then, a detailed relative examination is made with respect to different measures under several dimensions.

The chapter is organized as: some background information and related studies are given in Sections 8.2 and 8.3, respectively. The presented SS-KH algorithm is demonstrated in Section 8.4 and its results are analyzed in Section 8.5. Conclusions are derived in Section 8.6.

8.2 RESEARCH BACKGROUND

Swarm intelligence defines cooperative intelligence [12]. Biologists and natural scientists are continually examining the nature of social insects because of their effectiveness in solving the difficult problems like determining the shortest route from nest to food source or constructing nests [13, 14]. Despite the fact that the behaviors of these insects are unrefined separately, they make wonders as a swarm by interaction with each other and their environment. Recently, the nature of different swarms, which are utilized in the identification of prey or mating partner, is mimicked to numerical optimization technique [15]. In this study, unequal clustering problem has been resolved by the use of swarm intelligence techniques.

Energy-balanced unequal clustering (EBUC) [16] makes use of particle swarm optimization (PSO) algorithm to produce clusters of uneven sizes. It partitions the network into various parts of different sizes where the clusters near to BS be supposed to be of smaller sizes. The CHs near to BS save high energy which can be useful for data transmission between clustering and thereby hot spot problem will be eliminated. Genetic Algorithm based Energy-Efficient Adaptive Clustering Hierarchical Protocol (GAEEP) [17] is presented to increase network lifetime as well as network stability by electing optimum sum of CHs and its location by the use of GA. The working of GA is divided to setup as well as steady state phases. The effectiveness of GAEEP is verified by comparing its results with LEACH [18], SEP, ERP, LEACH-GA along with DEU in both homogeneous as well as heterogeneous networks in terms of lifetime, average remaining energy, and throughput.

Unequal clustering by improved particle swarm optimization (IPSO) [19] is developed to eliminate hot spot issue and also to overcome the usual limitations of PSO. The existing modified PSO algorithm may provide better performance but it suffers from high algorithmic complexity or high computational cost. It usually operates in numerous rounds where every round begins amid a setup phase continued by a steady state phase. The simulation results verified that the limitations of PSO algorithm are overcome by IPSO algorithm and is proved in terms of the number of alive nodes in WSN.

Sink mobility-based EBUC (SMEBUC) [20] is introduced for the attainment of balanced energy utilization by the use of shuffled frog leaping algorithm (SFLA). SFLA is used for the election of CHs and to organize clusters of varying sizes by the consideration of remaining energy level in the sensor node. To minimize the rotation of CHs often, CHs work continuously to identify the exchange time of CHs as well as node weights. The greedy algorithm is employed to choose the finest relay node stuck between CHs and BS. In addition, mobile sinks are presented to conquer the hindrance of hot spot concern. The highlight of the SMEBUC algorithm is verified by the performance comparison with LEACH and EBUCP in terms of energy dissipation as well alive nodes.

Novel chemical reaction optimization-based unequal clustering and routing algorithm (nCROUCRA) is proposed. A new chemical reaction optimization (nCRO) is proposed for asymmetrical clustering as well as routing algorithms and is called as nCRO-UCRA. To obtain unequal clustering, an nCRO paradigm-based CH selection is done using a derived cost function. In addition, a routing strategy is also developed using nCRO algorithm. They are designed with the effective models of molecular structure encoding as well as novel potential energy functions. This algorithm [21] is implemented in different scenarios based on the sensor count and CHs count.

8.3 LITERATURE SURVEY

At present, diverse models have been introduced using optimization techniques to address the hot spot issue. Yuan et al. [22] developed a genetic algorithm-based clustering model which determines the CH count and their location for lessening the utilization of energy in WSN. The functioning of this technique takes place in diverse iterations where every iteration comprises a setup and steady stage. At the former stage, the BS finds out the CHs and position. In the latter stage, a path is derived from the node to BS. It enables a node to transmit data in a straightforward manner to BS when the distance to BS is lower compared to the distance to CHs.

Fan and Du [23] generated different-sized clusters with respect to remaining energy level. Is also chooses the CHs using SFLA. It operates in two levels namely cluster construction and communication. The choice of CHs takes place in the earlier level and greedy method based path identification takes place in the latter level. Gajjar et al. [24] presented a clustering model which involves three levels, namely setup, neighborhood discovery, and steady state levels. At the initial two levels, node classification takes place in various layers and packet broadcasting takes place for neighboring node identification. It makes use of nonpersistent carrier sense multiple access (CSMA) for accessing the channel. At the last level, the process of CH election, construction of the cluster, and communication takes place. It makes use of fuzzy logic for selecting the CHs and optimum path selection takes place using ant colony optimization (ACO) algorithm.

Sabor et al. [25] developed an unequal clustering model for determining the cluster size and multi-objective immune technique for producing a routing tree. The size of the clusters will be identified using remaining energy level and distance to BS. Therefore, various optimization algorithms are applied for the identification of CHs and determination the cluster sizes. Once the CHs are chosen, the cluster will be constructed. Some of the hybridizations of unequal clustering techniques are presented in WSN. An energy balancing method for each cluster is introduced in Ref. [26]. To achieve this, the clusters are formed and then CHs are allocated to it. So, three steps of clustering take place, namely construction of clusters, election of CHs, and communication. At the first step, Sierpinski triangle is applied for creating small-sized clusters for the nodes located closer to the BS. On the selection of the CHs, it assumes the distance, remaining energy level, and node degree. A voting mechanism [27] to select the CHs takes place using remaining energy level, topology, and communication power. But, the selection of CHs takes place in a distributed way. It suffers from the drawback that this process extents the lifetime of a WSN. A Social Spider-based Unequal Clustering Protocol (SSUCP) for WSN presented in Ref. [28] is used to select the CHs and cluster size effectively.

8.4 THE PROPOSED SS-KH ALGORITHM

The presented SS-KH algorithm takes place in two stages, namely TCH using SS algorithm and FCH using KH algorithm. The SS-KH algorithm operates in four main stages, namely initialization, TCH using SS algorithm, FCH using KH algorithm, and cluster construction. Once the sensor nodes undergo deployment, initialization phase will be executed. At this stage, the nodes gather information about their neighboring nodes and distance to BS. Later, the SS algorithm utilizes the nature of social spiders to select the TCH. Then, the TCH undergoes selection of FCH and its respective CHs using the KH algorithm. Finally, the chosen FCHs form the clusters in the network. The pseudo code of the SS-KH algorithm is shown in Algorithm 8.1 and the entire process is demonstrated in Figure 8.3.

8.4.1 SS-Based TCH Selection

The SS optimization algorithm is based on behavior of SS and it is employed for electing the TCHs. The sub-processes involved in this mechanism are given below.

8.4.1.1 Representation of Individual Spiders

An essential process lies in the consideration that the SS algorithm is the way of representing the spiders. A group of $\boldsymbol{k}$ cluster center point is indicated by each set defined by each spider that is the most favorable solutions to unequal clustering issue. For instance, let $\boldsymbol{x} = \{(\mathbf{10.5; 20.4}), (\mathbf{15.2; 25.0})\}$ is a spider containing $\boldsymbol{m} = \mathbf{2}$ cluster centers and has {(10.5; 20.4) and (15.2; 25.0)}, each center has a dimension of $\boldsymbol{n} = \mathbf{2}$. Each spider in initial population is generated through the consideration of $\boldsymbol{k}$ arbitrary points of the dataset provided where $\boldsymbol{m}$ indicates the cluster count.

8.4.1.2 Distance between Two Spiders

Since the spider undergoes shaping by the collection of cluster centers, it is needed to define the distance between two spiders and not by point set. Therefore, the distance between a pair of spiders is represented as the sum of Euclidean distance between its cluster centers. The spiders with minimal distance hold smaller cluster size. For instance, assume $\boldsymbol{a} = \left\{\left(\boldsymbol{a}_{x1};\, \boldsymbol{a}_{y1}\right), \left(\boldsymbol{a}_{x2};\boldsymbol{a}_{y2}\right)\right\}$ ***and*** $\boldsymbol{b} = \left\{\left(\boldsymbol{b}_{x1};\boldsymbol{b}_{y1}\right), \left(\boldsymbol{b}_{x2};\boldsymbol{b}_{y2}\mathbf{2}\right)\right\}$ are the pair of spiders that has $\boldsymbol{m} = \mathbf{2}$ clusters centers, with each center comprising a set of two dimensions. Next, the distance between the pair of spiders is represented by:

$$\boldsymbol{d}_{a,b} = \boldsymbol{d}\left(\left(\boldsymbol{a}_{x1};\boldsymbol{a}_{y1}\right),\left(\boldsymbol{b}_{x1};\boldsymbol{b}_{y1}\right)\right) + \boldsymbol{d}\left(\left(\boldsymbol{a}_{x2};\boldsymbol{a}_{y2}\right),\left(\boldsymbol{b}_{x2};\boldsymbol{b}_{y2}\right)\right) \tag{8.1}$$

Consider $d\left(\left(a_{x1};a_{y1}\right),\left(b_{x1};b_{y1}\right)\right)$ is the Euclidean distance between the centers $d\left(\left(a_{x1};a_{y1}\right)\right.$ and $\left.\left(b_{x1};b_{y1}\right)\right)$.

8.4.1.3 Fitness and Weight of a Spider

The fitness function of each spider is determined by the use of a metric $\boldsymbol{M}$, a pointer to indicate optimal solutions which can be generated. The main objective of SS algorithm is the reduction of the population fitness. Therefore, in the population, the spider that has less fitness is the best one that can be selected as TCH in the cluster. The spider $\boldsymbol{i}$ is defined as the weight and fitness which has negative correlation, and it is determined as follows:

$$w_i = \frac{worst_s - J\left(s_i\right)}{worst_s - best_s} \tag{8.2}$$

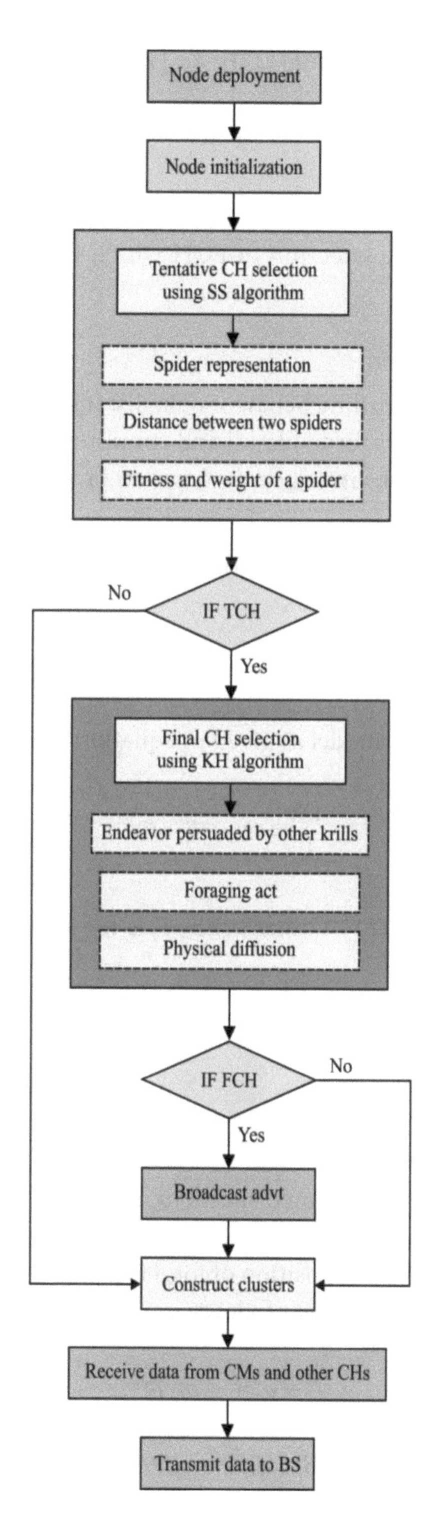

FIGURE 8.3 Overall process of proposed method.

$$best_s = \min\left(J\left(s_k\right)\right),\ k\epsilon\ \{1,2,\ldots,N\} \tag{8.3}$$

$$worst_s = \max\left(J\left(s_k\right)\right),\ k\epsilon\ \{1,2,\ldots,N\} \tag{8.4}$$

where $J\left(s_i\right)$ is the fitness rate of the spider. When the TCHs are selected, they will execute the KH algorithm for the selection of FCH which is discussed in the subsequent section.

8.4.2 KH-Based FCH Algorithm

KH algorithm is based on the herding behavior of krills. It is based on the krill individual's results. The group of krills hunts for foodstuff and communes with swarm members of the technique. A set of three motions in which the location of a krill has been presented:

- Endeavor persuaded by other krills
- Foraging act
- Physical diffusion

KH considered the Lagrangian model as given in equation (8.5).

$$\frac{dX_i}{dt} = N_i + F_i + Di \tag{8.5}$$

where N_i indicates the movement of other krills, F_i is the searching movement, and Di is the physical distribution.

In the initial movement, a target, local and repulsive outcome computes the direction of movement, α_i. For krill i, equation (8.6) is defined by:

$$N_i^{new} = N^{\max}\alpha_i + \omega_n N_i^{old} \tag{8.6}$$

where $N^{\max}$ indicates the utmost tempted speed, ω_n and N_i^{old} indicates the inertia weights and last motion respectively.

In the subsequent movement, the position of food is identified and earlier experience. At every ith krill, it could be represented as follows:

$$F_i = v_f \beta_i + \omega_f F_i^{old} \tag{8.7}$$

$$\beta_i = \beta_i^{food} + \beta_i^{best} \tag{8.8}$$

where $v_f, \omega_f, and\ F_i^{old}$ indicate the seeking speed, inertia, and final movement respectively, β_i^{food} indicates the attractiveness of food, and β_i^{best} indicates the consequence of optimal fitness of the ith krill.

In the final movement, an arbitrary procedure takes place in two ways, namely maximum diffusion speed and an arbitrary directional vector. Equation (8.9) is specified to:

$$D_i = D^{\max}\delta \tag{8.9}$$

where $D^{\max}$ and δ indicate the random vector. Using the set of three motions, the location of the krill at time t to $t+\Delta t$ is indicated as follows:

$$X_i(t+\Delta t) = X_i(t) + \Delta t \frac{dX_i}{dt} \tag{8.10}$$

The value of Δt is attained as follows:

$$\Delta t = C_t \sum_{j=1}^{NV} \left(UB_j - LB_j\right) \tag{8.11}$$

where NV represents the variable count, LB_j and UB_j are likewise the lower and upper bounds of the jth variables, C_t indicates a constant value of 0.5. The motion of the typical KH is influenced by other krills; for a preset number of generations or until termination condition is satisfied, foraging as well as physical diffusion prolongs to carry out. The transmission generally takes place in two ways, namely intracluster and intercluster communication. The data transmission takes place in a straight way or through intermittent CHs. The way of improving the data transmission within the cluster and the selection of proper CHs from every node in every iteration are the aim of clustering. The data from diverse CMs undergo aggregation at the CH and transmit the data to the BS. It got reduced in the quantity of the energy utilized in this method. At every iteration, every CH has to be allocated. A decision is made by the selection of appropriate node by the KH. An efficient CH in specific round is chosen based on the energy owned by node and distance from CHs member nodes that are not CH. At the setup stage, the sensors will relay the information related to the position and remaining energy level to the BS. The BS determines the average energy depending upon the data. At every iteration, the CH is chosen depending upon the maximum average energy in the specific iteration. Hence, a competitive node is chosen as the CH for that round. The BS undergoes the implementation of the algorithm to determine K number of proper CHs. It minimizes the cost function [26], as in equations (8.12)–(8.14):

$$\text{cost} = \beta \times f_1 + (1+\beta) \times f_2 \tag{8.12}$$

$$f_1 = max_{k=1,2,..k} \left\{ \sum_{\forall n_i \epsilon c_{p,k}} d\left(n_i, CH_{p,k}\right) / \left|C_{p,k}\right| \right. \tag{8.13}$$

$$f_2 = \sum_{i=1}^{N} E(n_i) / \sum_{k=1}^{k} E\left(CH_{p,k}\right) \tag{8.14}$$

where f_1 indicates the highest average Euclidean distance of nodes to their relevant CHs and $C_{p,k}$ indicates the node count fit to cluster C_k of krills. The function f_2 is represented as the ratio of sum of initialization energy of every node, $n_i = 1,2,.,N$, in the network with the totalized present energy of the CHs candidates in the present round. β is a predefined constant utilized for weighing the involvement of every subobjective. The intention of the fitness function is the minimization of intra-cluster distance among the nodes and CHs. It undergoes quantification by f_1. The quantification of energy efficiency takes place using f_2. Using the representation of cost function, lower values of f_1 and f_2 indicate that the cluster is optimum and has optimal node count. It also has the requirement of energy to perform the processes associated with the CH.

Step 1: Assume a collection of I krills which holds a set of K arbitrarily elected CHs among the appropriate CH candidates.

Step 2: Determine the cost function of every individual krill:

i. For every node $n_i = 1,2,.,N$

- Determine the distance $d(n_i,CH_{p,k})$ among the nodes n_i and every $CH_{p,k}$.
- Allocate the nodes n_i to $CH_{p,k}$ where:

$$d\left(n_i,CH_{p,k}\right) = f_1 = min_{\forall k=1,2,..k}\left\{d\left(n_i,CH_{p,k}\right)\right\} \tag{8.15}$$

ii. Compute the cost function by the use of equations (8.12)–(8.14)

Step 3: Determine the optimal value of every krill and identify the optimal position of the krills.

Step 4: Update the location of each krill in the search area by the use of following equations (8.16) and (8.17).

$$dXi = delta_t * \left(N(i) + F(i) + D(i)\right) \tag{8.16}$$

$$X(i) = X(i) + dXi \tag{8.17}$$

Step 5: Repeat steps 2–4 until the maximum number of rounds is obtained. The BS communicates the data containing the id of the CHs as well as CMs.

SS-KH FOR UNEQUAL CLUSTERING

INPUT:

Total Nodes ($Nodes_N$), Total Number of Clusters (C_N), $f()$

Initialize Population Size (Pop_N), Maximum number of Iterations (Max_{IT}), Number of Female Spiders ($Spider_F$), and Number of Male Spiders ($Spider_M$), Number of Krills (KH_N), Repulsive Effect (ω_N)

Initialize the iteration $t \leftarrow 1$

$$Pop_N \leftarrow Spider_M + Spider_F + KH_N$$

//Population Initialization

$\forall i \in Pop_N$ **do**

$$Pop_i \leftarrow diff\left(\left[0, sort\left(randperm\left(Nodes_N - 1, C_N - 1\right)\right), Nodes_N\right]\right);$$

end for

//Fitness Calculation

$\forall i \in Pop_N$ **do**

$$Fit_i \leftarrow f\left(Pop_i\right)$$

end for

repeat

//Calculation of Mating Radius

$\forall i \in Pop_N$ **do**

$$MR_i \leftarrow \frac{\sum_{e=1}^{Pop_M} m_e(o).W_{Pop_{F+h}}}{\sum_{e=1}^{Pop_M} Pop_{F+h}}$$

end for

//Calculation of weights of Each Spider

$\forall i \in Pop_N$ **do**

$$W_i = \frac{fit(P) - worst_{Pop}}{best_{Pop} - worst_{Pop}}$$

end for

//Calculation of Vibration of Female Spiders

$\forall i \in Spider_F$ **do**

$$Spider_F V_i = \begin{cases} Spider_F V_i(t) + x.V_{i,c}.\left(P_c - Spider_F V_i(t)\right) + y.V_{i,d}.\left(P_d - Spider_F V_i(t)\right) \\ +z.\left(rand - \frac{1}{2}\right) if\ r_l < \gamma \\ Spider_F V_i(t) - x.V_{i,c}.\left(P_c - Spider_F V_i(t)\right) - y.V_{i,d}.\left(P_d - Spider_F V_i(t)\right) \\ +z.\left(rand - \frac{1}{2}\right)\ if\ r_l \geq \gamma \end{cases}$$

end for
$\forall i \in Spider_M$ **do**

$$Spider_F V_i = \begin{cases} MV_i(t) + x.V_{i,f}.(P_f - Spider_M V_i(t)) + z.\left(rand - \frac{1}{2}\right) \\ if\ W_{Spider_{F+i}} > W_{Spider_{F+M}} \\ Spider_M V_i(t) + x.\left(\frac{\sum_{e=1}^{Spider_M} m_e(t).W_{Spider_{F+h}}}{\sum_{e=1}^{Spider_M} .W_{Spider_{F+h}}} - m_i(t)\right) \\ if\ W_{Spider_{F+i}} \le W_{Spider_{F+M}} \end{cases}$$

end for
//Impact of Krill-Herd
$\forall i \in KH_N$ **do**

$$\alpha_i = \alpha_i^L + \alpha_i^T // L \in Local\ Density, T \in Global\ Density$$

$$Pop_i^{new} = Pop^{max}\alpha_i + \omega_n Pop_i^{old}$$

end for
//Mating Process
// Cluster head Selection
$\forall j \in Pop_N$ **do**
$\forall j \in C_N$ **do**

$CH_i \leftarrow$ Choose a CH with Highest Residual Energy from $Pop_{i,j} \mid j \in C_N$

end for
end for
//Fitness Calculation

$$t \leftarrow t+1$$

until (termination condition satisfied)
OUTPUT: $min(f(Pop))$

8.5 EXPERIMENTAL VALIDATION

In this section, a detailed experimental validation takes place to verify the superior nature of the represented SS-KH algorithm under three unique scenarios. A set of evaluation parameters applied to measure the performance are energy efficiency and network lifetime.

8.5.1 Implementation Setup

The simulation takes place using MATLAB and a detailed experimental analysis is carried out with respect to three scenarios based on the distance to BS. To ensure the reliable

TABLE 8.1 Simulation Parameters

Parameters	Value
Area	100 × 100 m^2
Location (S1)	50, 50
Location (S2)	100, 100
Location (S3)	150, 50
E_0	0.5 J
Node count	200
E_{elec}	50 nJ/bit
ε_{fs}	10 pJ/bit/m^2
ε_{mp}	0.0013 pJ/bit/m^4
Packet size	4000 bits

performance of the SS-KH algorithm, results are measured based on diverse locations of BS which are mentioned as follows:

a. Scenario 1 (S1): BS at middle of the target area
b. Scenario 2 (S2): BS at corner of the target area
c. Scenario 3 (S3): BS located far away from the target area

A sample WSN consists of 200 nodes which undergo random deployment in the sensing field of 100 × 100 m^2. In addition, first-order radio energy model is considered. For comparative analysis, a set of five techniques, namely SS [28], LEACH [29], DEEC [30], TEEN [31], and FUCHAR [7], were taken. The assumption of the simulation parameters presented in the study is provided in Table 8.1. Besides, the deployment of nodes in all the three scenarios and the corresponding cluster construction is demonstrated in Figures 8.4 and 8.5 correspondingly.

8.5.2 Performance Analysis

8.5.2.1 Comparative Study on Energy Efficiency Analysis

To ensure the energy efficiency analysis of SS-KH algorithm, an investigation of average remaining energy takes place under three scenarios and the outcome is shown in Figures 8.6–8.8. The usage of energy is computed by the average energy consumed by every individual node. As seen in figure, it is evident that the introduced SS-KH algorithm shows more efficiency on comparing with other methods. It is due to the fact that efficient CH and proper cluster sizes are computed by the nature of SS-KH algorithm. The proper selection of CHs as well as size of the clusters minimizes the energy requirement in the entire WSN. In addition, the LEACH shows ineffective performance due to the nature of following characteristics: random CH selection and dedicated data transmission from CHs to BS. Simultaneously, TEEN offered minimum energy consumption over LEACH. However, it fails to achieve efficient results over the SS-KH algorithm. As TEEN is a threshold-based model, the decreased number of data transmission results in minimum energy utilization.

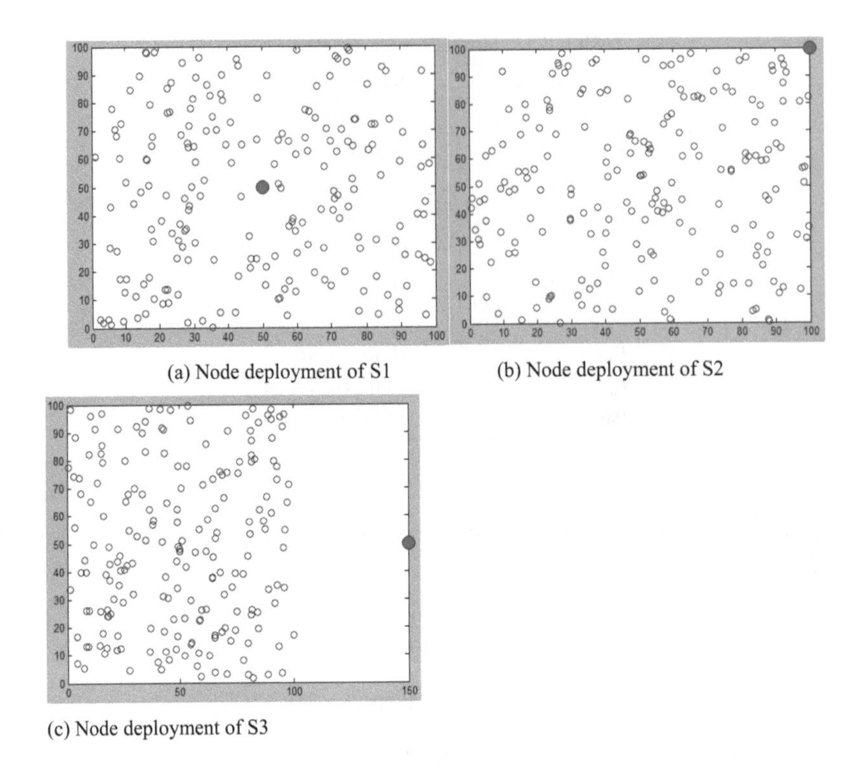

(a) Node deployment of S1 (b) Node deployment of S2

(c) Node deployment of S3

FIGURE 8.4 Deployment of nodes under three scenarios.

However, the way of selecting CHs randomly results in maximum energy expenditure over the compared techniques. The DEEC exhibits somewhat manageable performance over the other methods.

However, the selection of TCH arbitrarily results in poor performance compared to SS-KH, SSUCP, and *FUCHAR. Even though FUCHAR and SSUCP offer effective performance, they lead to inefficient outcome compared to the presented SS-KH algorithm. On all the applied different scenarios, the SS-KH algorithm showed extraordinary results over*

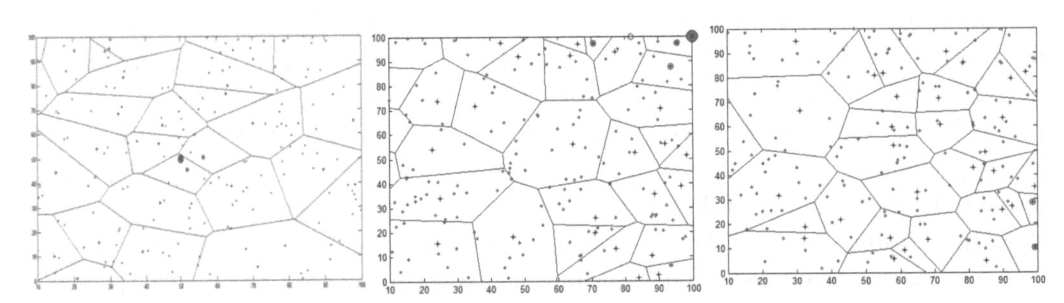

(a) Cluster construction of S1 (b) Cluster construction of S2 (c) Cluster construction of S3

FIGURE 8.5 Construction of clusters under three scenarios.

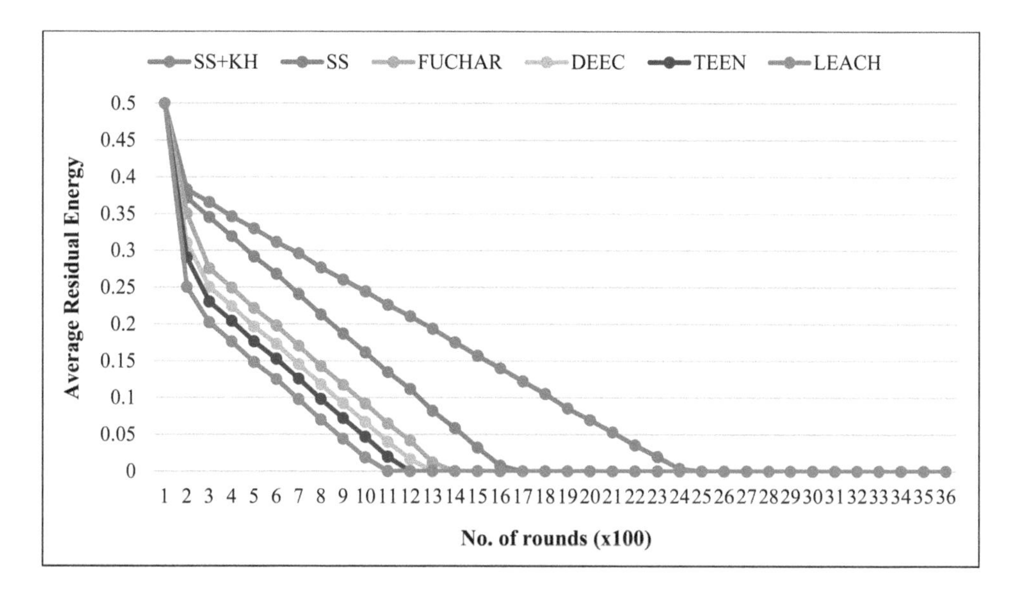

FIGURE 8.6 S1: Average energy consumption.

the methods used for comparison purposes. Once the data communication has begun, the remaining energy level starts to reduce and it stuck to null at certain time. In that case, the node will be declared as dead node and the alive node count begins to deceases. The clustering technique whose node count is high after the execution of numerous rounds is termed as an efficient model.

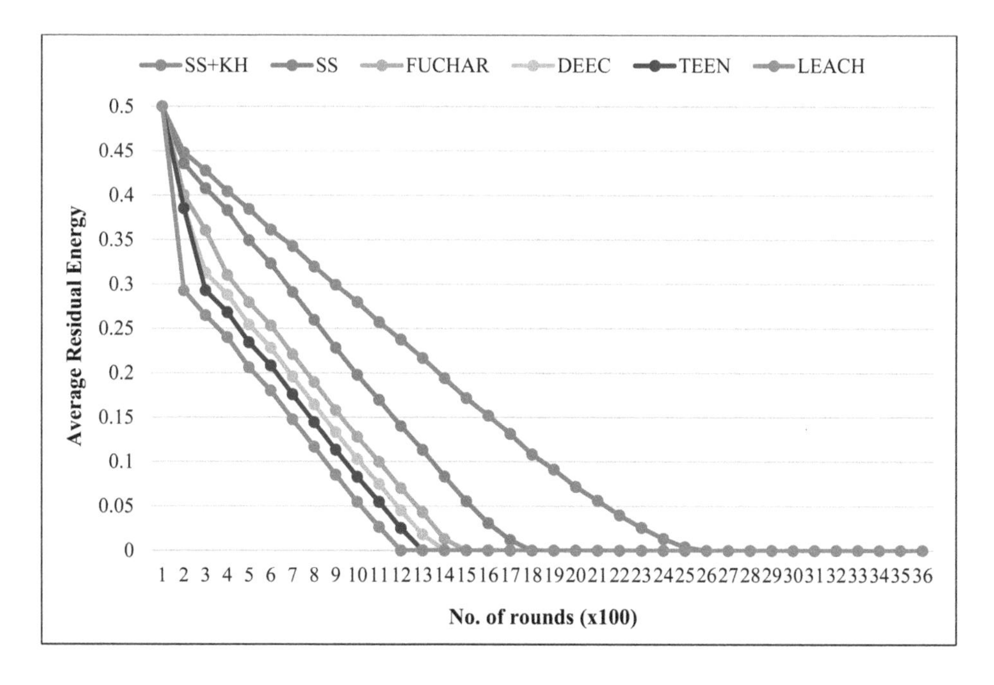

FIGURE 8.7 S2: Average energy consumption.

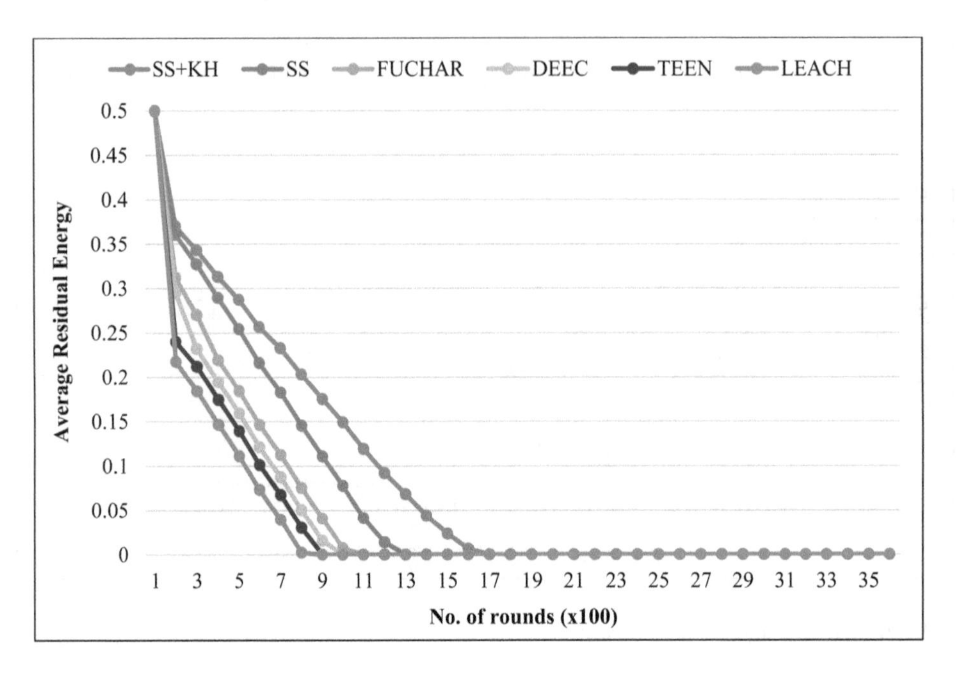

FIGURE 8.8 S3: Average energy consumption.

8.5.2.2 Comparative Study on Network Lifetime Analysis in Terms of Alive Nodes

From the various dimensions of validating the WSN lifetime, this study is based on the alive node count. From the view of deploying nodes, every node will stay alive. Figures 8.9–8.11 demonstrate the alive node count of diverse methods under three unique scenarios correspondingly. Similar to the average residual energy examination, minimum lifetime is

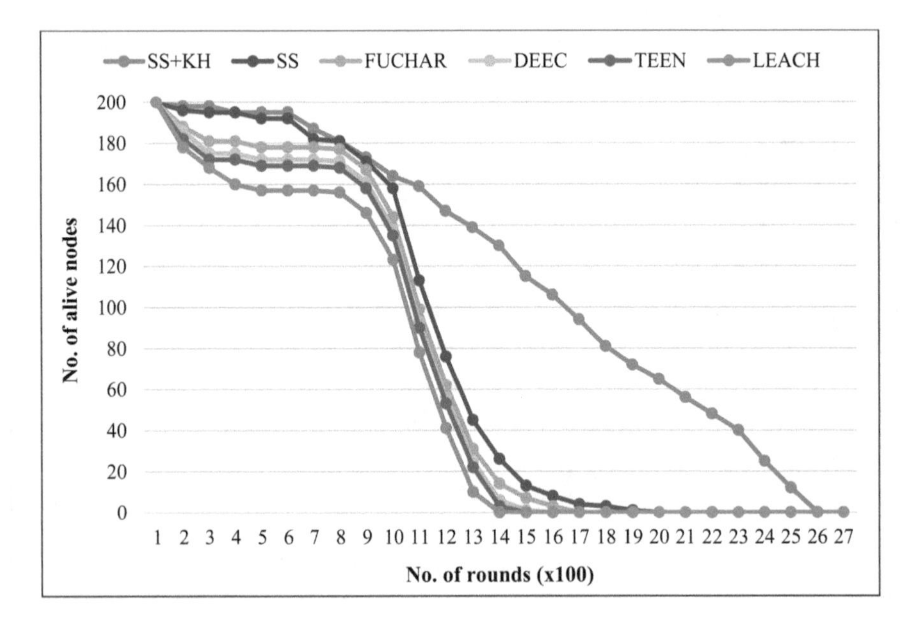

FIGURE 8.9 Network lifetime analysis of S1.

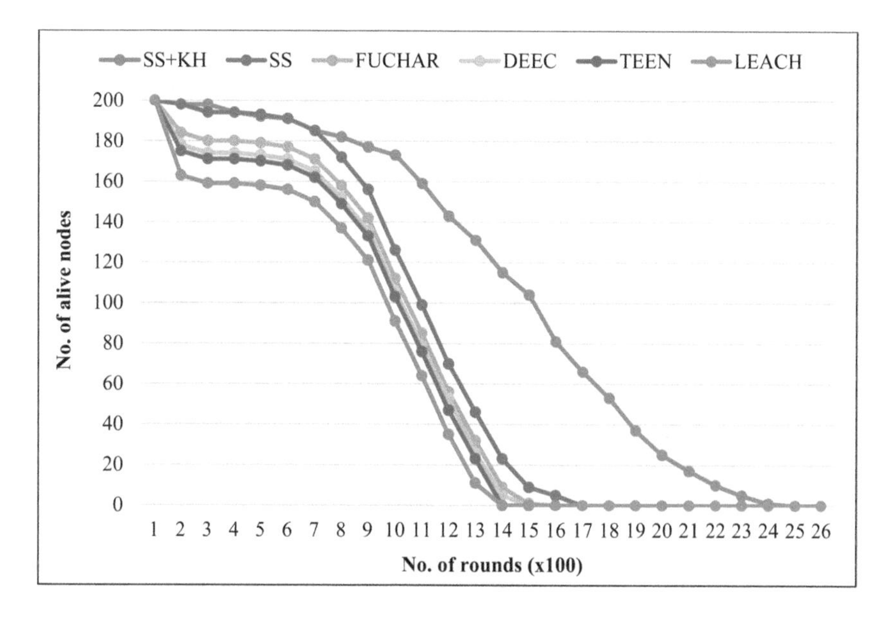

FIGURE 8.10 Network lifetime analysis of S2.

obtained by the LEACH over the compared methods. At the same time, TEEN performs well compared to LEACH, however, it failed to show better results over the other methods. In the same way, the DEEC, SSUCP, and FUCHAR are found to be effective over the compared techniques; however, the SS-KH is an effective technique. The figures clearly show that the maximum lifetime is attained by the SS-KH algorithm.

Figure 8.12 and Table 8.2 show the investigation of the results attained by different methods in terms of FND, HND, and LND. From the table, under the S1, it is clear that the

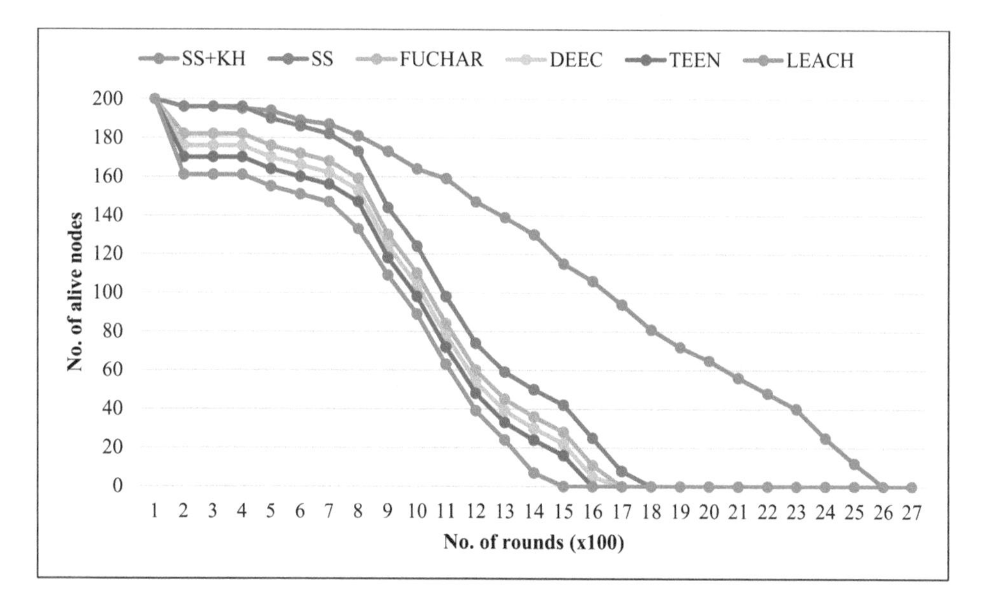

FIGURE 8.11 Network lifetime analysis of S3.

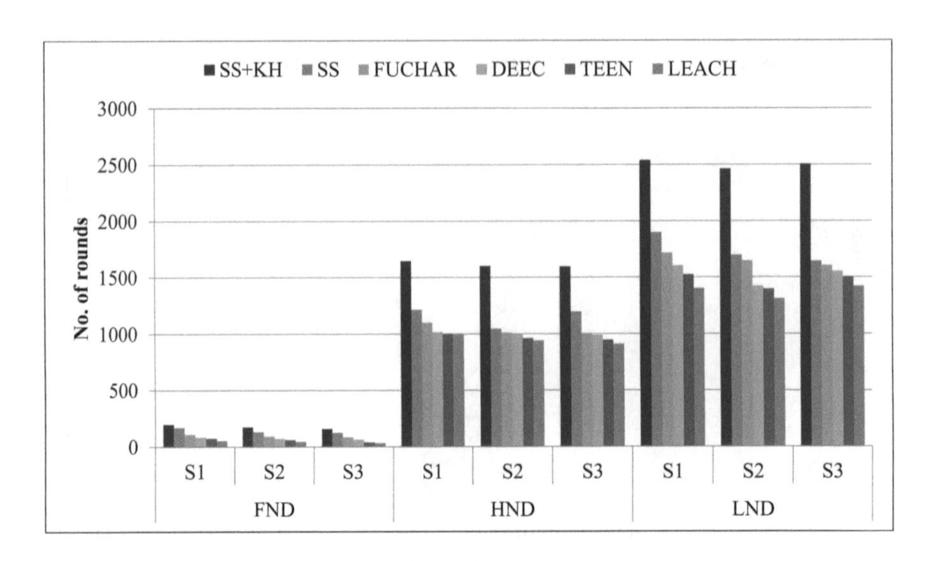

FIGURE 8.12 Network lifetime analysis in terms of FND, HND, and LND.

FND of the LEACH takes place by 53 rounds which its minimum network lifetime. Next, the TEEN provides better results than LEACH with the FND at 75 rounds. Simultaneously, the DEEC model somewhat lengthens the FND by 84 rounds. Next, the FUCHAR shows better lifetime with the FND of 110 rounds whereas the SS extends the FND to 170 rounds. However, the proposed SS-KH algorithm achieves maximum network lifetime with the FND of 198 rounds. Similarly, under S2 and S3, the proposed method achieves better performance in terms of FND. On measuring the results in terms of HND, under the S1, it is clear that the HND of the LEACH takes place by 996 rounds which its minimum network lifetime. Next, the TEEN provides better results than LEACH with the HND at 1001 rounds. Simultaneously, the DEEC model somewhat lengthens the HND by 1016 rounds. Next, the FUCHAR shows better lifetime with the HND of 1101 rounds whereas the SS extends the HND to 1046 rounds. However, the proposed SS-KH algorithm achieves maximum network lifetime with the HND of 1601 rounds. Similarly, under S2 and S3, the proposed method achieves better performance in terms of HND.

On measuring the results in terms of LND, under the S1, it is clear that the LND of the LEACH takes place by 1401 rounds which its minimum network lifetime. Next, the TEEN provides better results than LEACH with the LND at 1523 rounds. Simultaneously, the

TABLE 8.2 Comparison of Network Lifetime in Terms of FND, HND and LND

	FND			HND			LND		
Methods	**S1**	**S2**	**S3**	**S1**	**S2**	**S3**	**S1**	**S2**	**S3**
SS+KH	198	174	160	1645	1601	1595	2539	2460	2502
SS	170	131	124	1215	1046	1196	1900	1696	1642
FUCHAR	110	92	85	1101	1011	1006	1715	1646	1604
DEEC	84	72	65	1016	1001	989	1602	1421	1552
TEEN	75	60	40	1001	961	946	1523	1395	1501
LEACH	53	45	32	996	940	910	1401	1310	1420

DEEC model somewhat lengthens the LND by 1602 rounds. Next, the FUCHAR shows better lifetime with the LND of 1715 rounds whereas the SS extends the LND to 1900 rounds. However, the proposed SS-KH algorithm achieves maximum network lifetime with the LND of 2539 rounds. Similarly, under S2 and S3, the proposed method achieves better performance in terms of LND. These values proved that the SS-KH algorithm achieves maximum network lifetime over other methods under all the scenarios.

8.6 CONCLUSION

In this chapter, a new unequal clustering technique is presented with an intention of load balancing between the CHs with a view to avoid the hot spot problem. A new hybridization of SS and KH takes place to derive a new unequal clustering technique in WSN. The presented SS-KH algorithm takes place in two stages, namely TCH using SS algorithm and FCH using KH algorithm. Once the sensor nodes undergo deployment, initialization phase will be executed. At this stage, the nodes gather information about its neighboring nodes and distance to BS. Later, the SS algorithm utilizes the nature of social spiders to select the TCH. Then, the TCH undergo selection of FCH and its respective CHs using the KH algorithm. An extensive experimentation is carried out to verify the superior nature of the represented SS-KH algorithm under three unique scenarios. The simulation outcome depicted that the proposed model shows maximum network lifetime by lengthening the FND by 198, 1645, and 2539 rounds; HND by 174, 1601, and 2460 rounds; LND by 160, 1595, and 2502 rounds, respectively. These values verified the superior performance of the SS-KH algorithm. The SS-KH-UCP outperforms the compared methods LEACH, TEEN, FUCHAR, DEEC, and SSUCP in a significant way. In future, the proposed model can be implemented in real-time applications like smart home monitoring to control the home appliances, smart buildings, and smart cities.

REFERENCES

1. R. Buvanesvari and A. Rijuvana Begum, "A state of art approaches on unequal clustering in wireless sensor," *Int. J. Sci. Res. Rev.*, vol. 8, no. 3, pp. 719–729, 2019.
2. G. Chen, C. Li, M. Ye, and J. Wu, "An unequal cluster-based routing protocol in wireless sensor networks," *Wirel. Netw.*, vol. 15, no. 2, pp. 193–207, 2009.
3. T. Watteyne, "WSN Exploring Wireless Sensor Networking," International Workshop on Satellite and Space Communications, IEEE, Tuscany, Italy, 2009.
4. A. Sariga and P. Sujatha, "A survey on unequal clustering protocols in wireless sensor networks," *J. King Saud Univ.—Comput. Inf. Sci.*, vol. 31, pp. 304–317, 2019.
5. B. Baranidharan and B. Santhi, "DUCF: Distributed load balancing unequal clustering in wireless sensor networks using fuzzy approach," *Appl. Soft Comput. J.*, vol. 40, pp. 495–506, 2016.
6. J. Uthayakumar, T. Vengattaraman, and P. Dhavachelvan, "A new lossless neighborhood indexing sequence (NIS) algorithm for data compression in wireless sensor networks.," *Ad Hoc Netw.*, vol. 83, pp. 149–157, 2019.
7. S. Arjunan and P. Sujatha, "Lifetime maximization of wireless sensor network using fuzzy based unequal clustering and ACO based routing hybrid protocol," *Appl. Intell.*, vol. 48, pp. 2229–2246, 2018.

8. Y. Zhang, J. Wang, D. Han, H. Wu, and R. Zhou, "Fuzzy-logic based distributed energy-efficient clustering algorithm for wireless sensor networks," *Sensors*, vol. 1, 2017.
9. N. Mazumdar and H. Om, "Coverage-aware unequal clustering algorithm for wireless sensor networks," *Procedia Comput. Sci.*, vol. 57, pp. 660–669, 2015.
10. T. Liu, Q. Li, and P. Liang, "An energy-balancing clustering approach for gradient-based routing in wireless sensor networks," *Comput. Commun.*, vol. 35, no. 17, pp. 2150–2161, 2012.
11. R. Logambigai and A. Kannan, "Fuzzy logic based unequal clustering for wireless sensor networks," *Wirel. Netw*, vol. 22, no. 3, pp. 945–957, 2016.
12. J. Kennedy, R. Eberhart, and Y. Shi, *Swarm Intelligence*. London: Academic Press, 2001.
13. E. Bonabeau, M. Dorigo, and G. Theraulaz, *Swarm Intelligence: From Natural to Artificial Systems*. New York, NY: Oxford University Press, 1999.
14. M. Dorigo and L. M. Gambardella, "Ant colony system: a cooperative learning approach to the traveling salesman problem," *IEEE Trans. Evol. Comput.*, vol. 1, no. 1, pp. 53–66, 1997.
15. Z. Molay, R. Akbari, M. Shokouhifar, and F. Safaei, "Swarm intelligence based fuzzy routing protocol for clustered wireless sensor networks," *Expert Syst. Appl.*, vol. 55, pp. 313–328, 2016.
16. C. J. Jiang, W. R. Shi, M. Xiang, and X. L. Tang, "Energy-balanced unequal clustering protocol for wireless sensor networks," *J. China Univ. Posts Telecommun.*, vol. 17, no. 4, pp. 94–99, 2010.
17. M. Abo-zahhad, S. M. Ahmed, and N. Sabor, "A new energy-efficient adaptive clustering protocol based on genetic algorithm for improving the lifetime and the stable period of wireless sensor networks," *IJKESDP*, vol. 5, no. 3, pp. 47–72, 2014.
18. W. R. Heinzelman, A. Chandrakasan, and H. Balakrishnan, "Energy-efficient communication protocol for wireless microsensor networks," *Proceedings of the 33rd Annual Hawaii International Conference on System Sciences (HICSS)*, Maui, pp. 3005–3014, 2000.
19. S. Salehian and S. K. Subraminiam, "Unequal clustering by improved particle swarm optimization in wireless sensor network," *Procedia Comput. Sci.*, vol. 62, pp. 403–409, 2015.
20. M. E. Keskin, İ. K. Altıne, N. Aras, and C. Ersoy, "Wireless sensor network lifetime maximization by optimal sensor deployment, activity scheduling, data routing and sink mobility," *Ad Hoc Networks*, vol. 17, 2014.
21. S. Rao and Banka, "Novel chemical reaction optimization based unequal clustering and routing algorithm for wireless sensor networks (nCROUCRA)," *Wirel. Netw.*, vol. 23, no. 3, pp. 759–778, 2017.
22. A. M. R. Xiaohui Yuan, Mohamed Elhoseny, and Hamdy K El-Minir, "A genetic algorithm-based, dynamic clustering method towards improved WSN longevity," *J. Netw. Syst. Manag.*, vol. 25, no. 1, pp. 21–46, 2017.
23. X. Fan and F. Du, "Shuffled frog leaping algorithm based unequal clustering strategy for wireless sensor networks," *Appl. Math. Inf. Sci.*, vol. 9, no. 3, pp. 1415–1426, 2015.
24. S. Gajjar, M. Sarkar, and K. Dasgupta, "FAMACRO : fuzzy and ant colony optimization based MAC/Routing Cross-layer Protocol for wireless sensor networks," *Proc. Comput. Sci.*, pp. 235–247, 2014.
25. N. Sabor, Abo-Zahhad, S. Sasaki, and Ahmed, "An unequal multi-hop balanced immune clustering protocol for wireless sensor networks," *Appl. Soft Comput.*, vol. 43, pp. 372–389, 2016.
26. A. B. F. Guiloufi, N. Nasri, and A. Kachouri, "An energy-efficient unequal clustering algorithm using 'Sierpinski Triangle' for WSNs," *Wirel. Pers. Commun.*, vol. 88, no. 3, pp. 449–465, 2016.
27. H. Xia, R. Hua Zhang, J. Yu, and Z. Kuan Pan, "Energy-efficient routing algorithm based on unequal clustering and connected graph in wireless sensor networks," *Int. J. Wirel. Inf. Netw.*, vol. 23, no. 2, pp. 1–10, 2016.
28. R. Buvanesvari and A. Rijuvana Begum, "Social Spider based unequal clustering protocol for wireless sensor environment for smart cities," *Electron. Gov. Int. J.*, vol. 16, no. 1-2, pp. 190–209, 2020.

29. W. B. Heinzelman, A. P. Chandrakasan, and H. Balakrishnan, "An application-specific protocol architecture for wireless microsensor networks," *IEEE Trans. Wirel. Commun.*, vol. 1, no. 4, pp. 660–670, 2002.
30. S. Soro and W. B. Heinzelman, "Prolonging the lifetime of wireless sensor networks via unequal clustering," *Proceedings of 19th IEEE International Parallel Distributed Processing Symposium, IPDPS 2005*, vol. 2005, 2005.
31. A. Manjeshwar and D. P. Agrawal, "TEEN : a routing protocol for enhanced efficiency in wireless sensor networks," *Proceedings of 15th International Parallel and Distributed Processing Symposium, IPDPS 2001*, IEEE, San Francisco, CA, 2001.

CHAPTER 9

IoT Sensor Networks with 5G-Enabled Faster RCNN-Based Generative Adversarial Network Model for Face Sketch Synthesis

N. Gnanasankaran and E. Ramaraj

Department of Computer Science, Alagappa University, Karaikudi, Tamil Nadu, India

CONTENTS

9.1 INTRODUCTION

Nowadays, with progressive growth in the domain of IT sector and electronics, there has been an accelerated enhancement in the development of 5G IoT models. These techniques enable to brainstorm on the quality and efficiency of various city and residential facilities, including traffic management, energy, transportation, and so on. Simultaneously, cyber-crimes are considered as a crucial problem in daily life because they are not accumulated in a fixed local crime acts sets. Few possible traces of evidence are spread on several structures and cases, and they cross more jurisdictions ever before. Digital forensics is considered as a crucial part of the cybercrime investigation method. It is a technical attainment, research, and archive of data constituting the electrical broadcasting whose data could be used as proof in the law court. Forensic image processing contains the computer restoration and improvement of examination images [1]. It is focused on utilizing the data extraction from surveillance imaging, mainly for images that are noisy, incomplete, or over/underexposed. Forensic image technique is a method of improving the digital image with the help of various computer vision techniques. In computer imaging, face sketch synthesis (FSS) is an important zone, and it contains a broad range of applications like virtual social network and face recognition.

Assuming face identification as a tiny sample-sized issue that may be a real-time bottleneck, it may succeed to a great extent by creating training data with diverse differences. Face detection is an essential and significant issue in computer imaging and prototype identification that has been broadly considered over the past some years. It has been found useful in several face-related functions: face confirmation [2], face identification, face clustering, and so on. Subsequent to the new research of Viola Jones object recognition framework, several techniques have been planned toward face recognition in the past ten years. Earlier works in this field largely aimed at mining diverse kinds of handcrafted structures through field specialists in computer imaging and training-efficient classification to find and identify with conventional machine learning techniques. These methods are constrained in that they frequently need computer imaging professionals in creating better features and every specific element is enhanced particularly, creating the whole detection pipeline suboptimal. At present, deep learning (DL), particularly the deep convolutional neural networks (DCNNs), has attained outstanding success in many computer images processing, extending from image classifiers for detecting objects to semantic segmentation, and so on.

In contrast to the conventional computer vision technique, DL techniques avoid the handcrafted model pipeline subjected to several popular benchmark estimations, like ImageNet Large Scale Visual Recognition Challenge (ILSVRC) [3]. DL has been included in the zone of computer vision, which has led to the growth of research for exploring DL to resolve face detection process. Generally, face detection is considered as an unusual class of object detection method. Thus, researchers have tried to handle face detection by discovering effective DL methods for a basic object detection process. Most essential and extremely effective background for basic object recognition is the region-based CNN (RCNN) technique that is a type of CNN expansion to solve the object recognition process. Recent variations in face detections have enabled tracking regularly the research concepts through the extension of RCNN and its enhanced variations.

Although the technical development in face recognition exists, the creation of facial images with diverse adjustments and preserving at the same time the main identity is complex. In addition, there is a trouble in plotting from a dissimilarity factor to high-dimensional face imaging. It is also difficult that the human is extremely adapted to slight data in the facial region. In FSS, a set of two important challenges exist. Simultaneously, diverse variations in the semantic face exist such as expression, pose, facial disguise, and lighting changes that are difficult to synthesize within the image space.

For real-image space, a method for learning a complex modification effectually suitable to latent semantic space is yet a disputed thing.

Though the face recognition technique has been designed, the method of conserving the *subject identification* is a tough task. FSS has motivated several researchers to concentrate on this trendy subject. Creation of *cross-modal* images is a single-directional research area. The technique of Bayesian FSS presented in ref. [4] is the process of segmentation done with the help of two processes: weight calculation and neighbor selection technique. To accelerate the synthesis process, a new method of model-driven FSS is planned, and many effectual methods are discovered for enhancing the neighbor selection procedure. A method definitely used is repairing the obstructed portions of the facial images and performing appearance calculation of other facial parts by using a morphable method. For improvising the morphable technique process, a frontal face image dictionary at the initial stage is utilized [5]. This technique is planned based on principal component analysis (PCA) and verifies a goal image by linear eigenfaces combination. Sparse coding and cascaded pose regression are discovered for the age groups with many expressions and posture types. Due to artefacts, insufficient details of a particular facial feature, and resolution, the image quality formed by the actual techniques are minimized. It is difficult for the actual technique to identify facial images in their entirety.

Face recognition has been broadly investigated in the computer vision literature. Prior to 2000, despite numerous varied researches, the real-world efficient face recognition was far from reaching acceptable results until the revolutionary efforts made by Viola and Jones. Particularly, the VJ model [6] was the initial one to employ rectangular Haar-like features in a cascaded Adaboost classification to achieve actual face recognition. But it has many crucial disadvantages. Initially, its feature size was quite big. Classically, in a 24 × 24 deduction window, Haar-like features count was 160,000. Additionally, it is unable to efficiently deal with non-frontal and frontal faces in the wild. For resolving the initial issue, several works have come up with high complex features such as HOG, SIFT, SURF, and ACF. For instance, in ref. [7], a novel feature known as NPD was planned that calculated a ratio of variance between any two pixel intensity to the total values. Other authors focused on accelerating the feature selection process using heuristic techniques.

The famous Dlib C++ Library [8] has applied SVM as the classification technique in the face detection technique. Other techniques like random forest have also been tried. Improving the strength of recognition was another topic of research. One easy step is to integrate many detectors that had been developed distinctly for unique viewpoints or postures. Zhu and Ramanan [9] implied many deformable part techniques for capturing faces with diverse appearances. Shen et al. [10] suggested a retrieval-based technique associated with discriminative learning. However, training as well as testing these techniques consumes more time, and the improvement in recognition performance was comparatively restricted.

In recent times, Chen et al. [11] designed a technique for performing face detection concurrently with facial configuration, and attained maximum efficiency in accuracy as well as speed. Currently, they have observed the developments of face detection by DL, which repeatedly perform well in conventional computer vision technique. For instance, Zhan et al. [12] utilized CNN for automated learning and feature extraction process employed in face recognition. Li et al. [13] recommended a technique to detect faces in the wild that combines ConvNet and 3D mean face method in an end-to-end multitask discriminative learning background. Nowadays, Faster R-CNN [14] is utilized, which is one of the classical object detectors that has attained guaranteed outcomes. Moreover, many efforts have been carried out to enhance the Faster R-CNN model. In ref. [15], combined training performed on CNN cascade, region proposal network (RPN), and Faster R-CNN have been implemented in end-to-end optimization.

Lu et al. [16] showed the feasible method that includes two stages: preprocessing and sketch synthesis. Extensive surveys on open face sketch database approve that the existing plan enhances the sketch synthesis nature of the exemplar-based plans. Ye et al. [17] had proposed the triple interpretation GAN (TTGAN) with multilayer sparse illustration. They designed a multilayer scanty illustration technique, in which L1-standard illustration restrictions were combined with the image generation for improving the volume of character protection and the power of the created facial imaging for reconstruction fault. The face synthesis investigative outcomes on the standard face databases properly confirmed the effective performance with existing methods. In ref. [18], the CNN (M-CNN) technique has been improved, including two convolutional layers: a pooling layer and a multilayer perceptron (MLP) convolutional layer for learning the plotting among face images to sketches. However, the efficiency of this technique could be enhanced with the use of parameter optimization models.

This chapter introduces a new Internet of Things (IoT) and 5G-enabled Faster RCNN with generative adversarial network (GAN) called FRCNN-GAN model for FSS. The proposed model initially involves the image capturing process using IoT devices connected to the OpenMV Cam M7 Smart Vision Camera. It is used to capture the people faces from public places. Then, the Faster RCNN model is utilized for face recognition from the captured images. Next, the GAN model synthesizes the face images and generates the face sketch. Finally, the generated face sketch and the sketches that exist in the forensic databases are compared and the most relevant image is identified. A detailed experimental analysis indicated that the FRCNN-GAN model has been tested using two datasets: CUHK and IIIT. The simulation results indicated the effective performance of the presented model in terms of diverse measures.

9.2 THE PROPOSED FRCNN-GAN MODEL

The working principle involved in the presented FRCNN-GAN model is shown in Figure 9.1.

Initially, the FRCNN-GAN model acquires the images from public places by the use of OpenMV Cam M7 Smart Vision Camera. It captures the images and stores it in the memory. Then, the FRCNN-GAN model executes the face recognition process using Faster RCNN model, which identifies the faces properly in the captured image. Then, the GAN-based FSS module is employed to synthesize the recognized face and generate the face sketch. Finally, the generated face sketch and the sketches that exist in the forensic databases are compared and the most relevant image is identified.

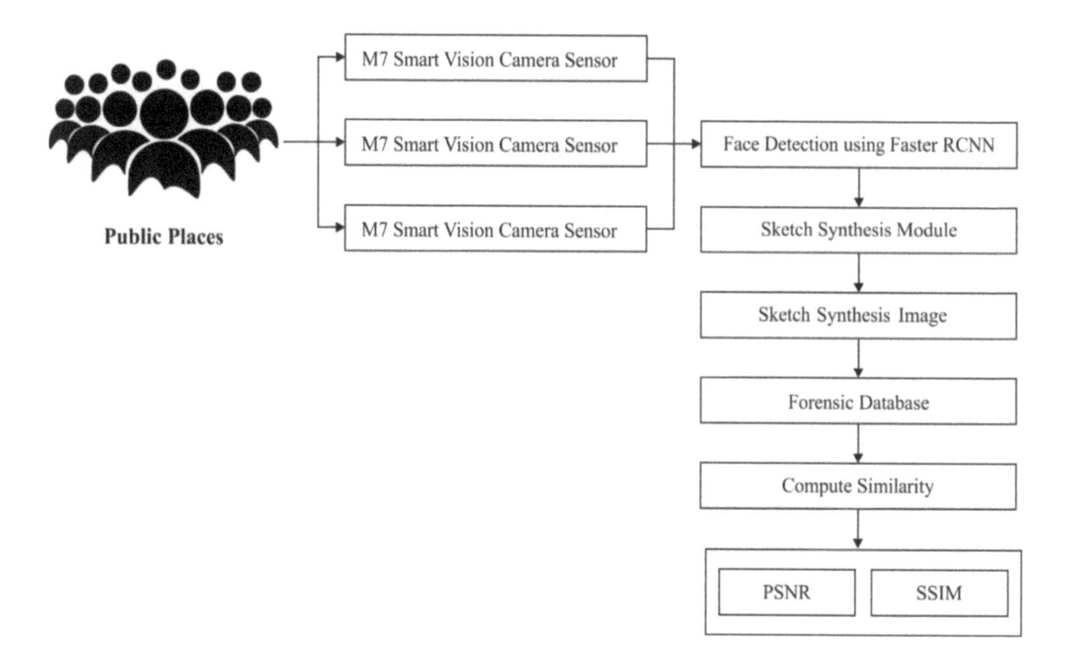

FIGURE 9.1 Block diagram of FRCNN-GAN model.

9.2.1 Data Collection

At the data collection stage, the proposed method makes use of 5G-enabled IoT devices called OpenMV Cam M7 Smart Vision Camera for data collection purposes. It comprises an OV7725 image sensing able to capture images at 640 × 480 8-bit greyscale or 320 × 240 16-bit RGB565 images at 30 FPS.

It involves an OpenMV camera that has a 2.8-mm lens on a standard M12 lens mount. It comprises a microSD card socket of 100 Mbs for read or write purposes. The SPI bus runs up to 54 Mbs and allows easy streaming of the image data. The sample image is depicted in Figure 9.2.

FIGURE 9.2 OpenMV Cam M7 smart vision camera.

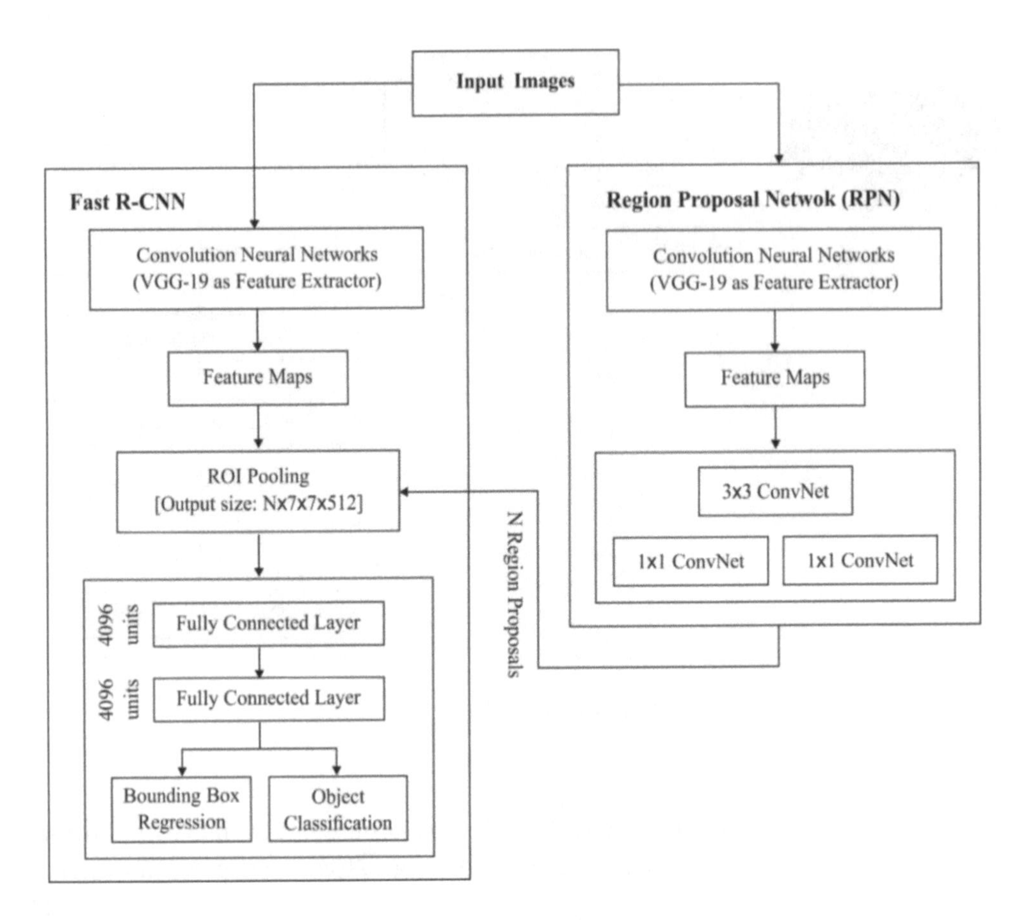

FIGURE 9.3 Overall architecture of Faster RCNN model.

9.2.2 Faster R-CNN-Based Face Recognition

It is an extremely upgraded version of R-CNN that is quicker and highly accurate in processing. The main alteration of Faster R-CNN is to utilize CNN for generating the object proposal in place of Selective Search in the earlier phase. It is known as *RPN*. At the higher level, RPN initially implied a base CNN network VGG-19 for extracting the features from the images. RPN yields image feature map as an input and makes a collection of object schemes sets, respectively, with an object score value as outcome. The minor network allocates object classifier scores sets and bouncing boxes directly to every object position. Figure 9.3 shows the overall structural design of Faster RCNN model. The steps involved in the Faster R-CNN are as follows:

- An image is taken and passed into the VGG-19 and the feature map as output for the image is obtained. RPN is employed on the feature maps. It is returned to the object proposals, including the object score.
- An RoI pooling layer is utilized in this method for reducing each proposal to the similar size.

- Finally, the approaches are used in a fully connected (FC) layer. It contains a softmax layer and linear regression layer at its top for classifying and resultant the bounding boxes to objects.

The RPN begins with the input image provided in the base of the CNN. The applied image initially resized to the smallest stride is 600 px through the larger stride not beyond 1,000 px. The outcome characteristics of the backbone network are generally shorter than the applied image based on the step of the backbone network. The feasible backbone network utilized in this effort is VGG16. It indicates that two successive pixels in the backbone outcome features signify two points of 16 pixels separately in the applied image. For every point in the feature map, the network learns whether an object exists in the applied image in the respective location and determines the size of the object. It can be performed by positioning "Anchors" sets on the applied image to every location on the outcome feature map from the backbone network. Such anchors stipulate probable objects in different sizes and feature ratios at this place. In total, nine feasible anchors in three dissimilar feature ratios and three various sizes are positioned on the applied image at point A on the outcome feature map. Anchors utilized have three scales of box regions 128^2, 256^2, and 512^2 and three aspect ratios of 1:1, 1:2, and 2:1.

As the network travels by every pixel in the outcome feature map, it verifies whether such k respective anchors crossing the applied image essentially have objects, and improving these anchors help attaining bound boxes as "Object proposals" or area of interest. Initially, a 3 × 3 convolutional layer with 512 units is used on the backbone feature map to provide a 512-d feature map to each location. It can be followed by two familial layers: a 1 × 1 convolutional layer with 18 units to object classifiers, and a 1 × 1 convolutional with 36 units to bounded box regressor. The 18 units in the classifier division provide an outcome with size (H, W, 18). These outcomes are utilized to offer a possibility of all points in the backbone feature map that comprises object inside every nine of the anchors at that time. The 36 units in the regression portion are applied to offer the four regression coefficients of every nine anchors for each point in the backbone feature map. These regression coefficients are utilized for enhancing the anchors that comprise objects.

- An anchor is considered "negative" when its IoU with each ground-truth boxes is lesser than 0.3. A residual anchor (either positive or negative) is ignored to RPN trained.

- A training loss to the RPN is a multitask loss provided by:

$$L(\{p_i\},\{t_i\}) = \frac{1}{N_{cls}} \sum_i L_{cls}(p_i, p_i^*) + \lambda \frac{1}{N_{reg}} \sum_i p_i^* L_{reg}(t_i, t_i^*) \tag{9.1}$$

- Here i is the index of the anchor in the mini-batch. The classifier loss $L_{cls}(p_i, p_i^*)$ is the log loss above two class labels (object versus not object). p_i is the outcome score from the classifier branch to anchor i, and p_i^* is the ground-truth label (1 or 0).

- The regression loss $L_{re}(t_i, t_i^*)$ is stimulated entirely and the anchor comprises an object, that is the ground truth p_i^* is 1. The word t_i is the outcome forecast of the regression layer and contains four variables $[t_x, t_y, t_w, t_h]$.
- The regression coefficients are employed for the anchors to accurate localization and offer proper bounded boxes.
- Each box is sorted as per their *cls* scores. Next, nonmaximum suppression (NMS) is utilized with a 0.7 as threshold value. The top-down bounded boxes that contain an IoU of higher than 0.7 with one or more bounding box are ignored. Therefore, the maximum-score bounded box is taken to the overlap box group.

The Fast R-CNN contains the CNN (usually pretrained on the ImageNet classifier task) with its last pooling layer exchanged through "ROI pooling" layer and its last FC layer is swapped by two separations—a $(K + 1)$ category softmax layer branch and a category-specific bounding box regression branch.

- The applied images are initially sent via the backbone of CNN for generating the feature map. In addition to the test time effectiveness, one more main purpose is to use an RPN as a proposal generator. It offers benefits of *weight distributing among the RPN and Fast R-CNN detector backbones.*
- Then, the bounding box approaches from the RPN are applied for pooling features from the backbone feature map. It can be performed via ROI pooling layer. An ROI pooling layer processes by (a) captivating the area equivalent for a method in the backbone feature map; (b) separating these areas into a static sub-windows count; and (c) executing max-pooling on this sub-windows for providing a static size outcome.

RoI pooling is a neural net layer utilized for object detection process. It was initially recommended by Ross Girshick in April, 2015. *It is a process of detecting objects by widely applying CNN. Its aim is to carry out max-pooling on the input of unusual sizes for obtaining fixed-size feature maps (e.g. 7×7).* It has accelerated the training as well as the testing process. It manages maximum detection accuracy. The results from the RoI pooling layer obtain a size of $(N, 7, 7, 512)$, where N is the approaches count from the RP technique. Subsequent to sending it to the two FC layers, a feature is provided into the sibling classifier and regression branches. It is noticeable that the classifiers and detection divisions are not similar to RPN. Here, the classifier layer has C units in all classes in the detection task. A feature is sent to a softmax layer for attaining the classifiers scores—the possibility of a suggestion related to every class.

9.2.3 GAN-Based Synthesis Process

Initially, the notations to the FSS are defined. Provide a test (observed) image t, the objective is generating the result s taking on M pairs of train face sketches and photos. The conditional GAN studies a nonlinear mapping from test image t and arbitrary noise vector z, for the result s, $\mathcal{G}:\{t, z\} \rightarrow s$ rather than $\{z\} \rightarrow s$ as GAN does. A generator $\mathcal{G}$ is studied for generating the results that could not be decided from "real" images by a discriminator $\mathcal{D}$ that is train for differentiating the generator's "fakes".

The objective of conditional GAN is written as follows:

$$\mathcal{G}^* = \arg\min_{\mathcal{G}}\max_{\mathcal{D}} \mathcal{L}_{cGAN}(\mathcal{G},\mathcal{D}) + \lambda\mathcal{L}_{L1}(\mathcal{G}) \tag{9.2}$$

where λ is for balancing the GAN loss as well as the regularization loss and the GAN loss is determined as follows:

$$\mathcal{L}_{cGAN}(\mathcal{G},\mathcal{D}) = \mathbb{E}_{t,s\sim p_{data}(t,s)}\left[log\mathcal{D}(t,s)\right] + \mathbb{E}_{t\sim p_{data}(t),z\sim p_z(z)}\left[\log\left(1-\mathcal{D}\left(t,\mathcal{G}(t,z)\right)\right)\right] \tag{9.3}$$

The conditional GAN loss is utilized for encouraging less blurring and is represented as follows:

$$\mathcal{L}_{L1}(\mathcal{G}) = \mathbb{E}_{t,s\sim p_{data}(t,s),z\sim p_z(z)\left[\|s-\mathcal{G}(t,z)\|_1\right]} \tag{9.4}$$

It is adapted to the generator as well as discriminator structures from individuals in the type of convolutional-Batch Norm-ReLu.

In this sketch synthesized by GAN, it maintains fine texture. But the noise appears with the fine texture because of the pixel-to-pixel mapping. For removing this noise, the sketch synthesized s and placed back onto the training sketches. Each face image is arranged and cropped for the identical size (250 × 200) based on the eye centers as well as mouth center.

Assume $X_1, \ldots, X_M$ indicates M training sketches. Initially, every training sketch and the sketch s are split into patches (patch size: p) through an overlapping (overlapping size: o) among neighboring patches. Assume $s_{i,j}$ signifies the (i, j)th patch from s, where $1 \le i \le R, 1 \le j \le C$. Now, R and C refer to the patch count in the path of rows and columns correspondingly to an image. As the sketch synthesized s has extremely same texture as that by training sketches, it has recreated the sketch s in a data-driven approach based on the Euclidean distance of image patches.

To sketch a patch $s_{i,j}$, it initially explores the K closer neighbors from every training sketch $X_1, \ldots, X_M$ around the location (i, j) with respect to their Euclidean distance among patch intensities. As there is disarrangement among various face sketches, it widens the explore area based on their respective place (i, j) by l pixels about its top, bottom, left, and right directions. So, it is $(2l + 1) \times (2l + 1)$ patches on all training sketches to match. To sketch patches $s_{i,j}$, it chooses K candidate neighbors from each $M(2l + 1)^2$ training sketch patch, indicated as $X_{i,j}^1, \ldots, X_{i,j}^K$. The recreation method is written as an easy regularization linear least-squares formulation as given in equation (9.5):

$$\min_{W_{i,j}} \left\| s_{i,j} - \sum_{k=1}^{K} W_{i,j}^k X_{i,j}^k \right\|_2^2 \tag{9.5}$$

$$s.t.\sum_{k=1}^{K} W_{i,j}^{k} = 1$$

where $W_{i,j} = \left(W_{i,j}^{1},\ldots,W_{i,j}^{K}\right)^{T}$ is the recreation weight. It has closed-form result as given in equation (9.6):

$$W'_{i,j} = \left(X_{i,j} - 1s_{i,j}^{T}\right)\left(\left(X_{i,j} - 1s_{i,j}^{T}\right)\right)^{T} / 1 \tag{9.6}$$

$$W^{(i,j)} = W'_{i,j} / \left(1^{T} W'_{i,j}\right)$$

where $X_{i,j} \in \mathbb{R}^{p^2 \times K}$ is the matrix of K neighbors and 1 is the vector of each 1*s*. It recreates the sketch patch $s_{i,j}$ as given in equation (9.7):

$$\hat{s}_{i,j} = \sum_{k=1}^{K} W_{i,j}^{k} X_{i,j}^{k} \tag{9.7}$$

Finally, each recreated patch $s_{i,j}(1 \le i \le R, 1 \le j \le C)$ is arranged into a complete sketch $\hat{s}$ through overlapping area average.

9.3 PERFORMANCE VALIDATION

A detailed experimental analysis takes place on two databases: IIT and CUHK. The sample set of images is shown in Figure 9.4.

The qualitative analysis of the FRCNN-GAN model is shown in Figure 9.5. As depicted, the figure showed that the FRCNN-GAN model has generated the sketch image highly resembling to the input image.

Table 9.1 and Figures 9.6 and 9.7 analyze the FSS results of the FRCNN-GAN model in terms of PSNR and SSIM on the applied two datasets.

Figure 9.6 shows the PSNR analysis of the FRCNN-GAN model on the applied two datasets. On the applied CUHK dataset, the experimental values indicated that the MWF

FIGURE 9.4 Sample images. (a) IIIT dataset. (b) CUHK dataset.

TABLE 9.1 Results Analysis of the FRCNN-GAN with Existing Methods in Terms of PSNR and SSIM

		Methods					
Dataset	**Measures**	**MRF**	**MWF**	**SRGS**	**SCDL**	**CNN**	**FRCNN-GAN**
CUHK	PSNR	15.07	14.41	14.79	15.14	15.64	16.12
	SSIM	0.58	0.59	0.58	0.59	0.59	0.62
IIIT	PSNR	19.26	17.20	18.46	18.33	19.62	20.23
	SSIM	0.54	0.57	0.59	0.58	0.61	0.65

and SRGS models have showed minimum PSNR values of 14.41 and 14.79 dB, respectively. At the same time, the MRF and SCDL models have resulted in slightly higher PSNR values of 15.07 and 15.14 dB, respectively. Besides, the CNN model has shown somewhat better performance with the high PSNR value of 15.64 dB. Furthermore, the FRCNN-GAN model has resulted in a higher PSNR value of 16.12 dB. On the given IIIT dataset, the experimental measures showcased that the MWF and SCDL frameworks have exhibited lower PSNR values of 17.20 and 18.33 dB, respectively. The SRGS and MRF approaches have shown better PSNR values of 18.46 and 19.26 dB, respectively. Also, the CNN method has shown considerable value with the high PSNR value of 19.62 dB. The FRCNN-GAN method has showcased to maximum PSNR value of 20.23 dB.

Figure 9.7 implies the SSIM analysis of the FRCNN-GAN method on the applied two datasets. On the provided CUHK dataset, the experimental scores pointed that the MRF and SRGS methodologies have showcased lower SSIM values of 0.58 and 0.58 dB, respectively. Simultaneously, the MWF, CNN, and SCDL schemes have attained better and same SSIM value of 0.59 dB. Moreover, the FRCNN-GAN approach has provided maximum SSIM value of 0.62 dB. On the applied IIIT dataset, the experimental values pointed that the MRF and MWF methodologies have pointed lower SSIM values of 0.54 and 0.57 dB, respectively. Concurrently, the SRGS and SCDL frameworks have attained moderate SSIM values of 0.59 and 0.58 dB, respectively. Additionally, the CNN technology has implied manageable performance with the best SSIM value of 0.61 dB. Also, the FRCNN-GAN model has provided greater SSIM value of 0.65 dB.

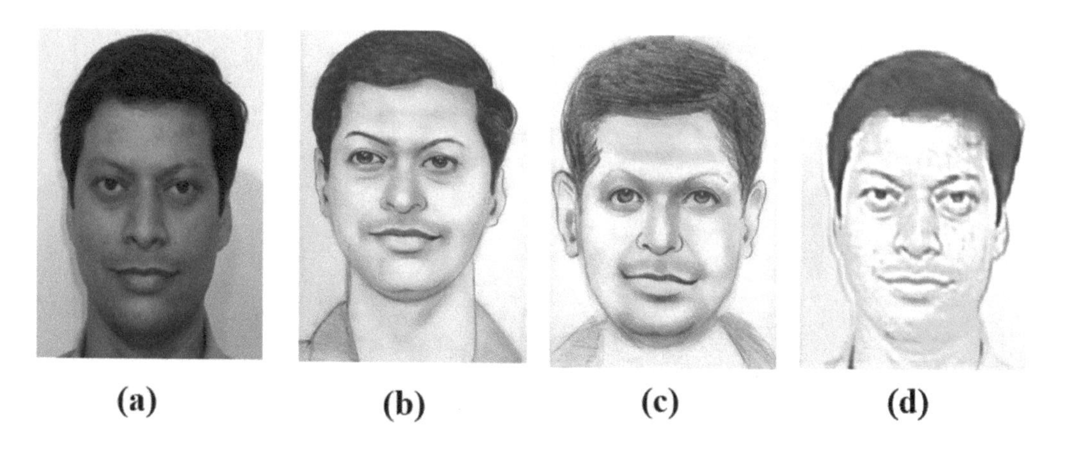

FIGURE 9.5 (a) Input image. (b) Viewed sketch. (c) Forensic image. (d) Sketch synthesis image.

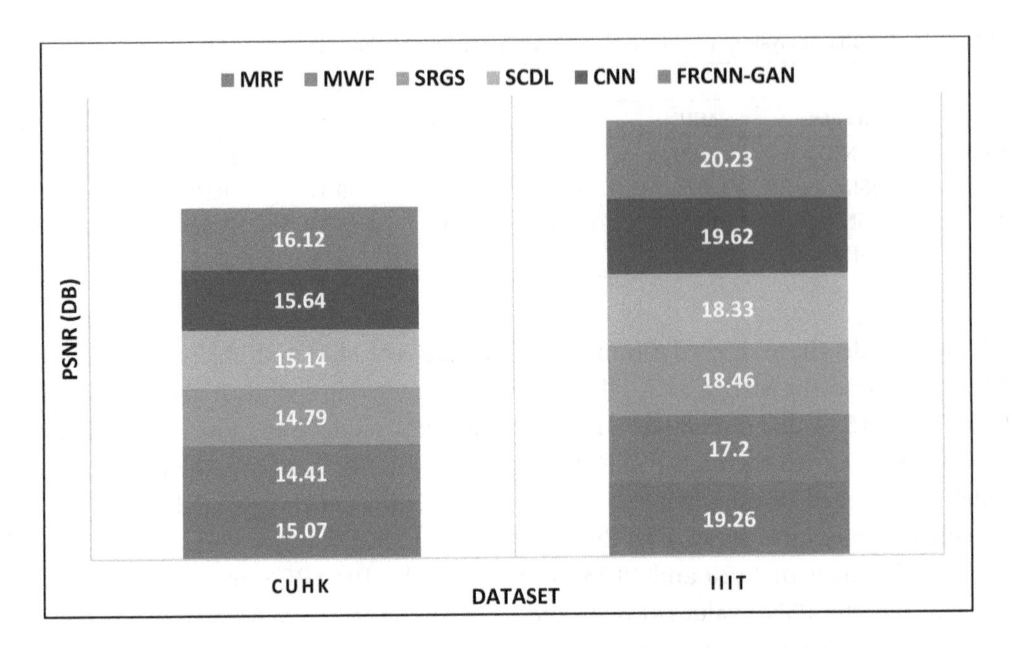

FIGURE 9.6 PSNR analysis of FRCNN-GAN model on the applied two datasets.

Table 9.2 and Figures 9.8 and 9.9 examine the accuracy analysis of the FRCNN-GAN approach on the applied two datasets. Figure 9.8 depicts the accuracy analysis of the FRCNN-GAN method on the applied two datasets. On the applied CUHK dataset, the experimental measures notified that the SCDL and MWF methodologies have exhibited lower accuracy values of 69.85% and 70.84%, respectively. Simultaneously, the MRF and

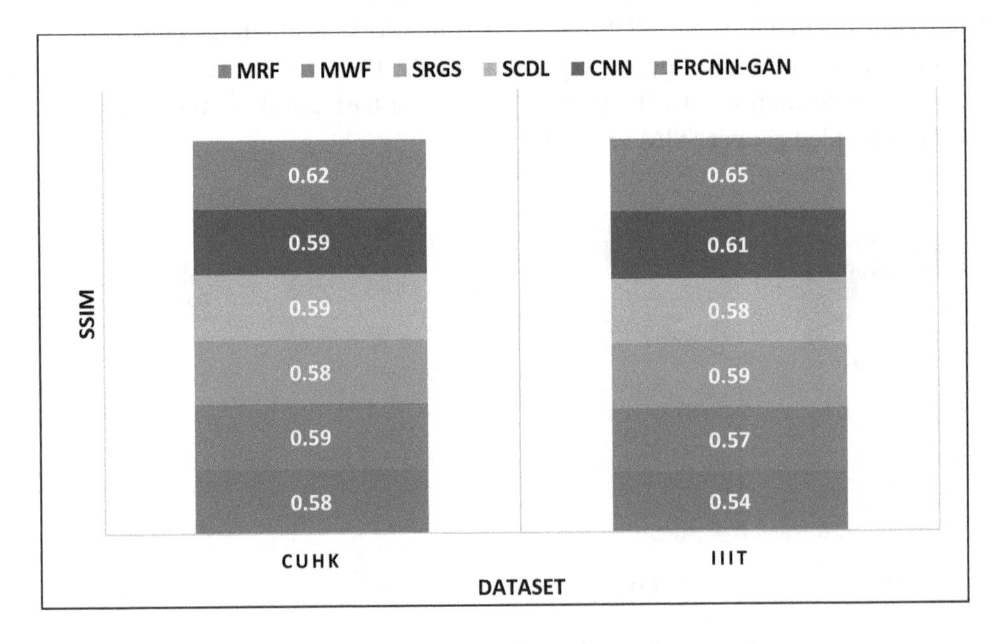

FIGURE 9.7 SSIM analysis of FRCNN-GAN model on the applied two datasets.

TABLE 9.2 Accuracy Analysis of the FRCNN-GAN with Existing Methods.

Methods	CUHK	IIIT	Average
MRF	71.30	71.34	71.32
MWF	70.84	68.30	69.57
SRGS	72.45	72.40	72.43
SCDL	69.85	71.75	70.80
CNN	78.53	80.21	79.37
FRCNN-GAN	80.56	81.40	80.98

SRGS approaches have offered reasonable accuracy values of 71.30% and 72.45%, respectively. Furthermore, the CNN approach has exhibited moderate function with better accuracy value of 78.53%. Additionally, the FRCNN-GAN technique has concluded with maximum accuracy value of 80.56%. On the applied IIIT dataset, the experimental values pointed out that the MRF and MWF models have depicted lower accuracy values of 68.30% and 71.34%, respectively. Besides, the SCDL and SRGS methodologies have accomplished reasonable accuracy values of 71.75% and 72.40%, respectively. Additionally, the CNN approach has shown slightly better performance with maximum accuracy value of 80.21%. Moreover, the FRCNN-GAN approach has showcased a remarkable accuracy value of 81.40%.

Figure 9.9 displays the average analysis of the FRCNN-GAN approach on the applied two datasets. It states that the experimental values pointed out that the MWF and SCDL models have showcased least average values of 69.57% and 70.8%, respectively. Concurrently, the

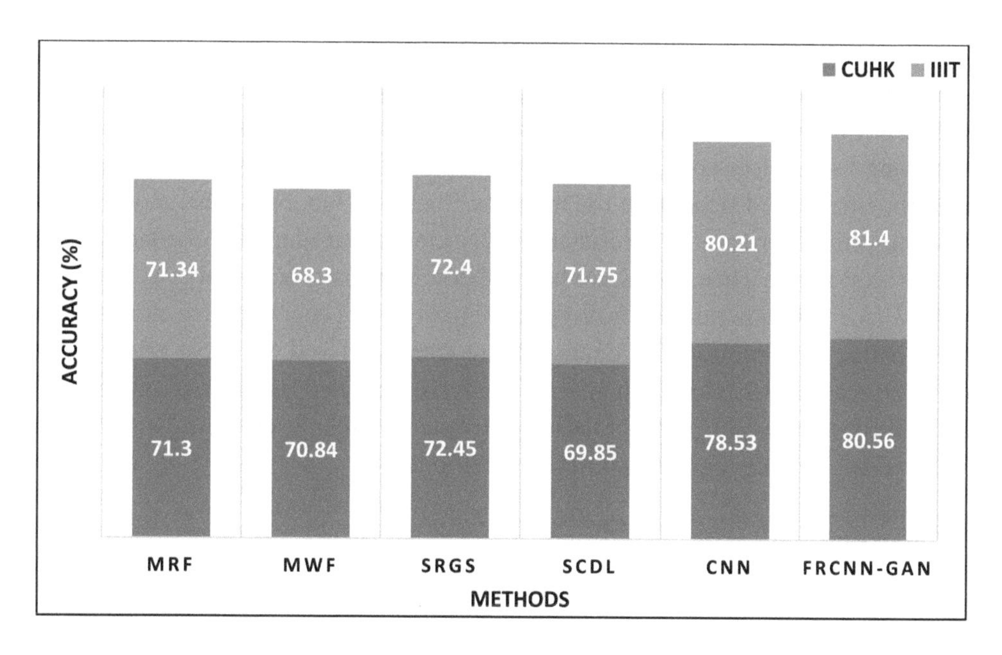

FIGURE 9.8 Accuracy analysis of FRCNN-GAN model on the applied two datasets.

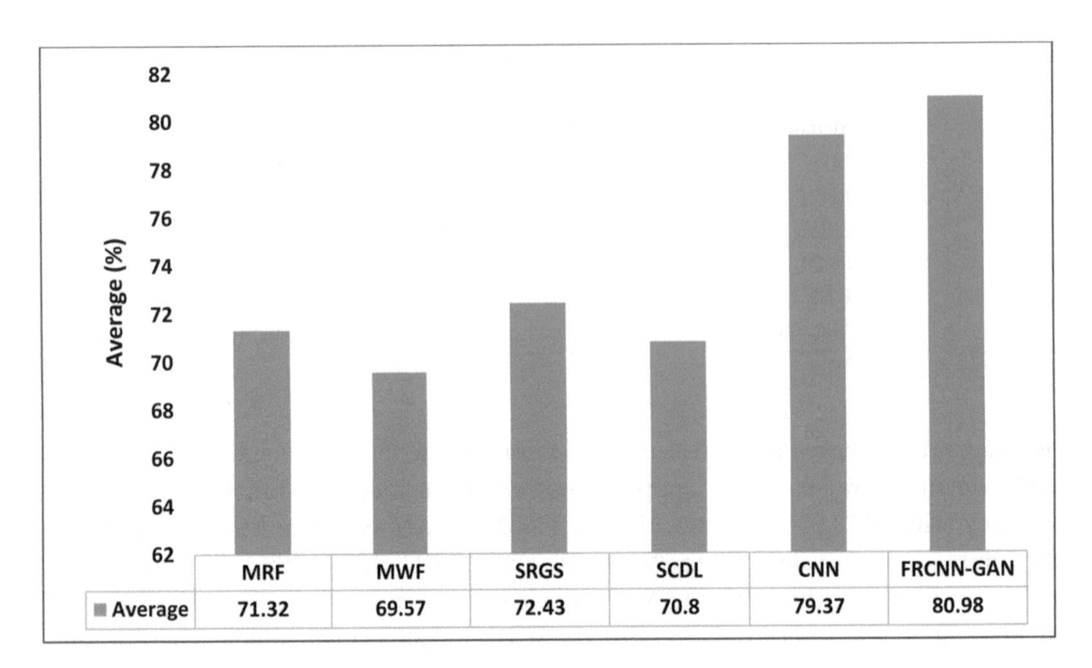

FIGURE 9.9 Average analysis of FRCNN-GAN model with existing methods.

MRF and SRGS models have concluded with better average values of 71.32% and 72.43%, respectively. Also, the CNN approach has depicted considerable function with the higher average value of 79.37%. Additionally, the FRCNN-GAN framework has resulted to a maximum average value of 80.98%.

9.4 CONCLUSION

This chapter has developed a new IoT and 5G-enabled Faster RCNN with GAN called FRCNN-GAN model for FSS. The proposed model initially involves the image capturing process using IoT devices connected to the OpenMV Cam M7 Smart Vision Camera. It captures the images and stores it in the memory. Then, the FRCNN-GAN model executes the face recognition process using Faster RCNN model that identifies the faces properly in the captured image. Then, the GAN-based FSS module is carried out to synthesize the recognized face and generate the face sketch. Finally, the generated face sketch and the sketches that exist in the forensic databases are compared and the most relevant image is identified. An extensive experimentation analysis is carried out on two databases: IIT dataset and CUHK dataset. The simulation outcome ensures the effective performance of the FRCNN-GAN model over the compared methods.

ACKNOWLEDGMENT

This research work was carried out with financial support of RUSA–Phase 2.0 grant sanctioned vide Letter No. F. 24-51/2014-U, Policy (TNMulti-Gen), Department of Education, Government of India, 9.10.2018.

REFERENCES

1. C.S. Brown, Investigating and prosecuting cyber crime: forensic dependencies and barriers to justice. *Int. J. Cyber Criminol.* 9 (1) (2015) 55.
2. Y. Sun, X. Wang, X. Tang, Deep learning face representation from predicting 10,000 classes, in: *Proceedings of the IEEE Conference on Computer Vision and Pattern Recognition*, 2014, pp. 1891–1898.
3. A. Krizhevsky, I. Sutskever, G.E. Hinton, ImageNet classification with deep convolutional neural networks, in: *Proceedings of the Advances in Neural Information Processing Systems*, 2012, pp. 1097–1105.
4. S. Klum, H. Han, A.K. Jain, B. Klare, Sketch based face recognition: forensic vs. composite sketches, in: *2013 International Conference on Biometrics (ICB)*, 2013, June, IEEE, pp. 1–8.
5. N. Wang, X. Gao, J. Li, Random sampling for fast face sketch synthesis, *Pattern Recognit.* 76 (2018) 215–227.
6. P. Viola, M. Jones, Rapid object detection using a boosted cascade of simple features, in: *Proceedings of the IEEE Conference on Computer Vision and Pattern Recognition (CVPR)*, IEEE, 1, 2001, pp. 511–518.
7. S. Liao, A.K. Jain, S.Z. Li, A fast and accurate unconstrained face detector, *IEEE Trans. Pattern Anal. Mach. Intell.* 38 (2) (2016) 211–223.
8. D.E. King, Dlib-ml: a machine learning toolkit, *J. Mach. Learn. Res.* 10 (2009) 1755–1758.
9. X. Zhu, D. Ramanan, Face detection, pose estimation, and landmark localization in the wild, *Proceedings of the IEEE Conference on Computer Vision and Pattern Recognition (CVPR)*, IEEE, 2012, pp. 2879–2886.
10. X. Shen, Z. Lin, J. Brandt, Y. Wu, Detecting and aligning faces by image retrieval, in: *Proceedings of the IEEE Conference on Computer Vision and Pattern Recognition*, 2013, pp. 3460–3467.
11. D. Chen, S. Ren, Y. Wei, X. Cao, J. Sun, Joint cascade face detection and alignment, in: *Proceedings of the European Conference on Computer Vision*, Springer, 2014, pp. 109–122.
12. S. Zhan, Q.Q. Tao, X.H. Li, Face detection using representation learning, *Neurocomputing* 187 (2016) 19–26.
13. Y. Li, B. Sun, T. Wu, Y. Wang, Face detection with end-to-end integration of a ConvNet and a 3D model, in: *European Conference on Computer Vision*, Springer, Cham, 2016, pp. 420–436.
14. H. Jiang, E. Learned-Miller, Face detection with the faster R-CNN, in: *Automatic Face & Gesture Recognition (FG 2017), 2017 12th IEEE International Conference on IEEE*, 2017, pp. 650–657.
15. H. Qin, J. Yan, X. Li, X. Hu, Joint training of cascaded CNN for face detection, in: *Proceedings of the IEEE Conference on Computer Vision and Pattern Recognition*, 2016, pp. 3456–3465.
16. D. Lu, Z. Chen, Q.J. Wu, X. Zhang, FCN based preprocessing for exemplar-based face sketch synthesis, *Neurocomputing* 365 (2019) 113–124.
17. L. Ye, B. Zhang, M. Yang, W. Lian, Triple-translation GAN with multilayer sparse representation for face image synthesis. *Neurocomputing* 358 (2019) 294–308.
18. L. Jiao, S. Zhang, L. Li, F. Liu, W. Ma, A modified convolutional neural network for face sketch synthesis, *Pattern Recognit.* 76 (2018) 125–136.

CHAPTER 10

Artificial Intelligence-Based Textual Cyberbullying Detection for Twitter Data Analysis in Cloud-Based Internet of Things

Srikanth Cherukuvada,[1] Kiranmai Bellam,[2] A. Soujanya,[3] and N. Krishnaraj[4]

[1]*Department of Computer Science and Engineering, Sasi Institute of Technology and Engineering, Tadepalligudem, Andhra Pradesh, India*

[2]*Department of Computer Science, Prairie View A & M University, Prairie View, TX, USA*

[3]*Department of Information and Communication Engineering, Anna University, Chennai, Tamil Nadu, India*

[4]*School of Computing, SRM Institute of Science and Technology, Kattankulathur, Tamil Nadu, India*

CONTENTS

10.1 INTRODUCTION

The instances of Internet harassment have become higher in the cyber world and can occur anytime during conversations in which people see, participate, or share content. Torture or provocation can be perceived as "sad behavior" with an intention of harming others. The occurrence of cybercrime in cyberspace includes erroneous messages, images, sounds, and videos that undermine or annoy [1]. Digital security is primarily aimed at young people and adolescents as it is the most dynamic form of meeting, using innovation, for various purposes such as participation, socialization, and so on. Cyberbullying involves sending malicious messages, defamatory comments, or images through unofficial communications, mail, and so on. The goals behind these are basically the parts of the web: tireless, the ability to watch and copy, just like the detectable crowd. Since the electronic exclusion significantly affects the society [2], it is highly important to consider this research field as the need of the hour [3]. Although this is a variant of harassment that occurs in offline world, the way it is done online has its own characteristics. For example, after the presentation of a demonstration on Internet, it could remain open on the Internet forever and not just breaking the normal barriers required for harassment [4]. Three strategies, such as the keyword system, mining estimation, and Internet life test are used in the methodology. These methods are combined as tests to deal with cyberbullying [5].

The online competition is divided into an initial section in which a short electronic avoidance database and all its basic procedures are provided following which a description of machine learning (ML) techniques is given. ML methods are used to identify the business on Internet. ML is characterized as the ability of a computer on decision-making with the help of accessible information and meetings [1]. The first step in understanding and preventing electronic blocking is its identification, and here we are likely to present potentially unsafe e-mails. These messages create particular difficulties for normal Natural Language Processing (NLP) as these messages are strange and full of spelling mistakes and outlines [6]. Percussion recognition strategies can be divided into two categories: one by slogan and another by Artificial Intelligence (AI). The easiest way is password technique that uses slogans to search for a sensitive substance in content. Although the thinking is basic, this technique is able to achieve high precision score using both search terms and web search [7]. The AI method is getting increasingly entangled. The three main sections of AI are representation, evaluation, and learning and these three comparative rules are designed for e-commerce recognition techniques. In the recognition of content based on impact circuit, the fundamental and foremost progress is the numerical representation of learning for immediate messages [8]. Indeed, the representation of content learning has largely focused on content extraction, data recovery, and NLP. Cyberbullying and cybercrime are interesting online topics to explore in detail [9].

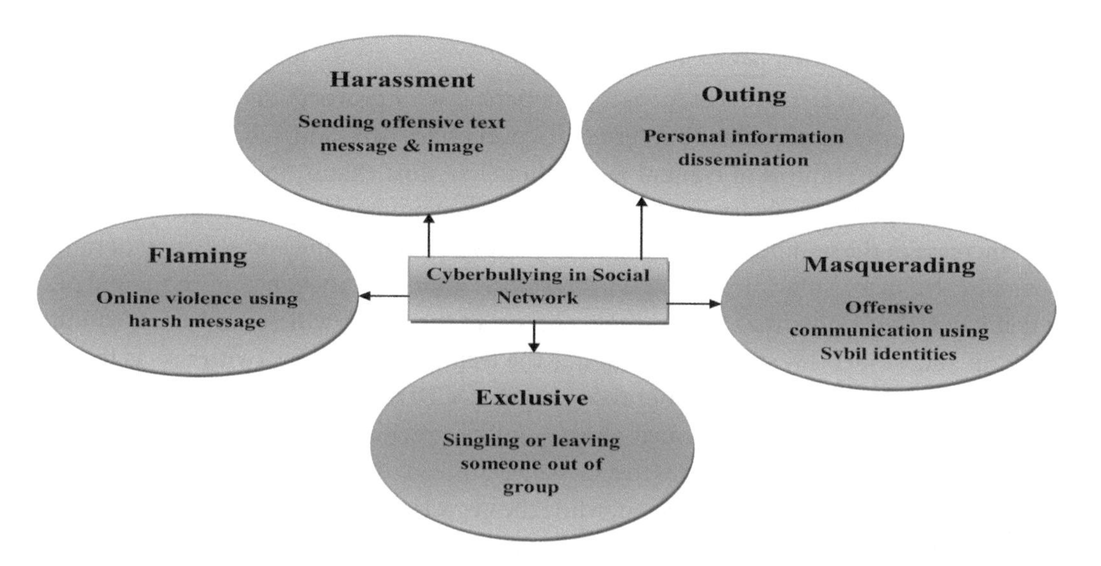

FIGURE 10.1 Cyberbullying types.

A huge taskforce focuses on identification and non-observation techniques to locate cyberbullying with the help of remarkable data from sexually explicit data, UI, and etymological and non-verbal references to realistic properties [10]. The discovery of digital death is considered as critical thinking and another calculation is suggested that aims at reducing the time needed to run an electronic promotion notice, thus limiting the amount of estimates required to select options. The planned position in cyberspace requires computational methodologies that can benefit from different properties of both vocal and non-verbal types [11]. In addition, several mechanical learning models are considered, including direct models, tree-based models, and deep learning models with numerous labels printed so as to make them an ideal model. Taking the collected tweets into account, the nature of planned discovery in digital madness is appreciated while the best model identified can reach an accuracy of over 90% with expected content expectations [12].

The levels of harassment have increased in India in manifolds. A total of 79% of Indians are aware and concerned about cyberbullying compared to 54% across the globe. A total of 53% Indians underwent cyberbullying, compared to normal 37% worldwide [13]. Likewise, half of the Indian population has involved in Internet harassment and only 24% of the total population has been confronted. In particular, the word Net database is used to identify semantically related words and to assess the proximity to selected cyberbullying terms. Again, a regulation-based methodology is proposed in the current study to identify the real cases of cyberbullying [14].

This chapter is organized as follows. Section 10.1 states the introduction phase in cyberbullying. Section 10.2 discusses the related works in cyberbullying. Then, Section 10.3 describes the proposed methodology in detail. Section 10.4 reports the experiments along with their results. Section 10.5 discusses the reported results, concludes, and summarizes this research chapter.

10.2 LITERATURE REVIEW

Online badgering is a crime that focuses on a person with provocation and contempt on Internet. Numerous approaches to Internet discovery have been adopted, though it largely depends on the usefulness of content and customers. Most exams, found in writing, are designed to improve discovery by introducing new key points. Despite the fact that there is an increase in the number of markings, the processes like sending and selection of functions become increasingly annoying. Furthermore, another annoyance of these updates is that it certainly underlines, for example, the age of the client, which can be easily built. Al-Ajlan and Ykhlef [15] suggested to update the technology used by Twitter so as to recognize e-commerce-based e-learning recognition (OCDD), which is one another way of addressing the challenges mentioned above. Unlike previous works, it doesn't separate skills from tweets and pass them on to a classifier. Instead, it talks to an escort as a whole. Word semantics are currently protected, while the negotiation periods and screening selections can be rejected. As far as grouping phase is concerned, in-depth learning is used in parallel with the modernization of calculation to modify the parameters.

Banerjee et al. [16] suggested that the progress is increasing rapidly. These ongoing advances changed the way people collaborate widely while sending mail, another measure. In any case, despite the way in which the evolution supports many parts of life, it is accompanied by different effects that somehow affect the people. The search for death is one of those results. Cyberbullying is a crime in which a culprit focuses on a person with an annoying online test that has serious emotional, social, and physical consequences for the victim. To address this problem, this study proposed another technique to distinguish digital mourning based on deep nervous system. The nervous system was used and better results were achieved than the existing frames.

With ever-growing progress in Internet, a corresponding immediate increase is also observed in the number of people using it for mailing purposes. As a result, cyberbullying has harmed the competitor, for example, electronic blocking, a type of abuse of others who use data innovation in a targeted and coherent way. Differentiation and prevention of cyberbullying is fundamental for normal phases of savings and prosperity. Mahlangu et al. [17] examined Internet content, categories of information on Internet, and sources of information containing information on electronic block for research and AI systems to distinguish online harassment. The main difficulties in tissue identification were observed, including the absence of media-based discoveries and the accessibility of information indicators available to the general population. The proposals were provided at the end of the audit.

Internet vaccination is a dangerous kind of mental abuse, as cyber victims, especially children and young people, suffer from the negative effects of emotional well-being problems that could lead to self-destructive thoughts. Further, Internet harassment is a serious problem in the Middle East region. The existing commitments to recognize cyberbullying focus mainly on English language. Aghbari and co-workers [18] introduced a continuous approach to address the discrimination based on cyberbullying in Arab Twitter feeds. Likewise, it organized the messages that harass users, based on their quality. If a "forward" message is detected, the box notifies the customer and suggests a move based on the strength of the torture message. The importance of the proposed approach was shown in this study by suggesting how it could be used by a parent to monitor their children's activities and warn them if any suspicious action is recognized. The research inferred that the proposed framework was able to practically discern the Internet messages in a consistent manner.

Nazar et al. [19] recommended that the issue of recognition of electronic bullying must be adequately addressed. As a clear difference from most of the past works that seek to discern strong behavior by considering individual messages, this study considered the electronic barrier as unnecessary compulsive behavior toward an individual and together many messages that got exchanged between customers were examined in this study. In addition, the researchers were ready to quickly choose an expensive option. To this end, another approach was proposed in this study with different levels. In the first level, (i) a solitary message is described as valid or not and the ideal and minimum number of data from that message is evaluated and (ii) this new information is used to choose whether to continue checking messages or end the process and run a digital alarm. The proposed approach seemed to ensure that most of the messages were detected before selecting an option, while the ideal selection principle demonstrated that it limits the normal Bayes. An authentic evaluation of the information on Instagram shows that the proposed technique was able to accurately identify the cases of cyberbullying, observing up to 59% less messages than innovation.

Cyberbullying is a new demo that annoys, humiliates, weakens the self or annoys others through electronic gadgets and online interpersonal websites. Online cyberbullying is more dangerous than ordinary harassment, as it can increase the embarrassment for an unlimited online crowd. According to UNICEF and a study conducted by the Ministry of Communications and Information, Indonesia, 58% of 435 young people do not understand the term "cyberbullying." Some of them may have been a threat; however, since they do not understand harassment in digital technology, they could not perceive the negative consequences of their torment. Threats may not perceive the damage done to their business, because they do not see the immediate reaction of their victims. Nurrahmi and Nurjanah [20] designed a methodology to distinguish cyberbullying artists based on their written work and to analyze the accuracy of customer information and inform them of the damage caused by cyberbullying. This research collected the information from Twitter. Since the information was not tagged, an online tagging device was created to characterize the tweets in cyberbullying and not cyberbullying. A total of 301 tweets about cyberbullying, 399 tweets without cyberbullying, 2,053 negative words, and 129 abusive words were collected from the device. So the authors used SVM and KNN to find out and identify the cyberbullying records. The results show that the SVM secured the highest score f1, 67%. In the same way, the customer confidence survey was conducted that found 257 regular customers, 45 harassing and evil characters on screen, 53 painful artists, and 6 painful screen characters on the screen.

In the study conducted by Pascucci et al. [21], it was planned to show the importance of supporting computational stylometry (CS) and ML to recognize the creator's age and gender in articles on cyberbullying. This study created a phase of localization of cyberbullying, terms of accuracy; revision and measurement of *F* for sexual orientation and determination of age in the works that we have collected take part in the performance.

10.3 PROPOSED METHODOLOGY

Various researchers proposed different methods of detecting cyberbullying; however, they were largely based on text and user-defined functions. Most of the studies found in the literature aim to improve the detection by introducing new features. However, with an increase in the number of elements, the selection and selection steps of the elements get complicated. An input dataset is sent for data preprocessing and is used to improve the quality of the input data. Data preprocessing also includes the removal of keywords and special characters. After preliminary data processing, the output data is sent to the feature extraction process where the optimal method is selected through classification. Then the oppositional grasshopper optimization with convolutional neural network (OGHOCNN) classification algorithm is used to detect the cyberbullying words in tweets. The key characteristics are determined using the reduction method and combined into a single model that offers the best detection accuracy. The proposed Twitter data based cyberbullying detection method shown in Figure 10.2.

10.3.1 Preprocessing

This section focuses specifically on the features built into the current study's cyber threat detection model. This includes user personalities that focus on Big Five, Dark Triad models, Twitter-based emotions, passions, and features. Social network data is serious due to which preprocessing is applied to improve the accuracy of input data. This includes the removal of stop words. Stop words are generally "a," "like," "have," "is," "the," "o," etc., which are nothing but "Stop using words," a phenomenon important for memory space and processing time.

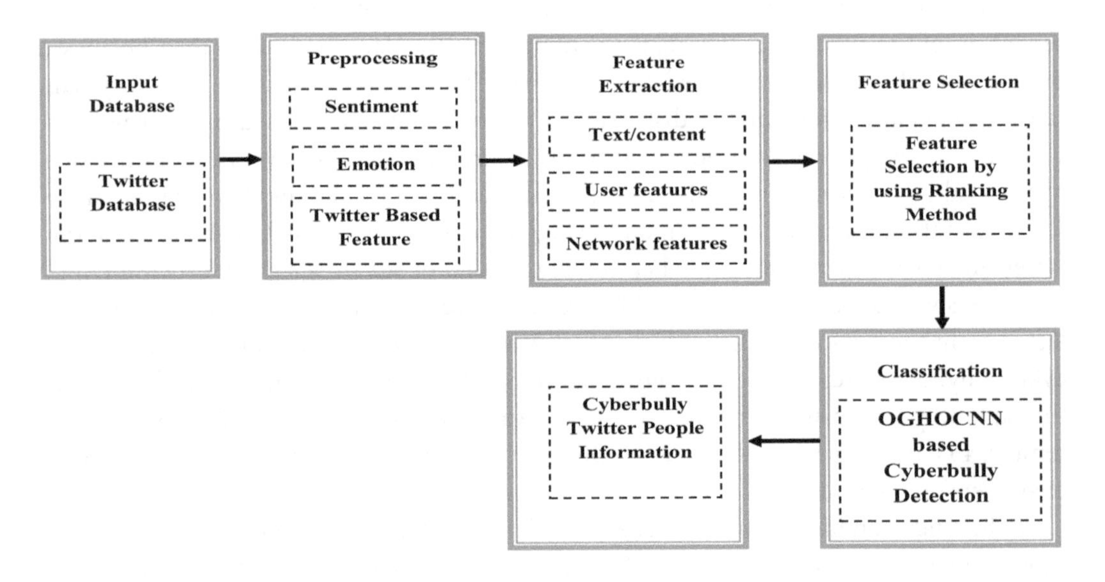

FIGURE 10.2 Proposed twitter data-based cyberbullying detection.

10.3.2 Feature Extraction

Twitter includes Twitter API as extracted to identify digital vulnerabilities, while additional coding has been created to remove highlights, such as lowercase and uppercase letters (i.e., content features). The highlights removed are:

- ***Text/content features***—including font sizes, uppercase, lowercase, hash tags, images, client references, URLs, and multimedia content.
- ***User features***—functionality removed from each person's profile, for example, age by age, number of status checks (i.e., the number of tweets—number of retweets performed by a customer), number of registrations (i.e., open registrations that are part of a customer), and the number of customer's best choices (a customer with his number) bill greeted by the amount of tweets (Chatzakou et al., 2017a).
- ***Network features***—Customer-specific steps involve social measurement such as the amount of followers, devotees, and their reputation (the extent of support for supporters).

10.3.3 Feature Selection Using Ranking Method

Based on a supervised methodology, tweets can be obtained as a basic unit or a modified one to find the average scores for a language. The estimate mainly shows how common is the amount of irritating words in the progress of tweets, when deciding the progress of previous tweets. When a tweet contains quick words, they are available at that time, whereas the level of importance determines the scores. It is now indistinguishable from the inclination of the multinomial network, in which all the tweets of a class are converted into individual tweets. After this process, the feasibility of these classy tweets is assessed. To the extent of administrative significance, the parameter class addresses the correct tweets and also the full set of readiness. It is recognized that a part shows the time in dataset and the time in class tweets. The length of the information collection (e.g., the planning set) is class and individual based on the frequency of the words collected. The data shown in Figure 10.1 denotes the length of the set and the speed of the class. The glory of false alarms is taken into account and is defined in equation (10.4).

$$P = \sum_{d \in s} \sum_{w \in d} tf_w \tag{10.1}$$

$$Q = \sum_{d \in c_j} \sum_{w \in d} tf_w \tag{10.2}$$

$$R = \frac{L}{M} \tag{10.3}$$

$$NFA\left(w, a_j, s\right) = \binom{k}{m} \cdot \frac{1}{R^{m-1}} \tag{10.4}$$

The meaning of phrase win set j c is as follows:

$$Meaning\left(w,a_j\right)=-\frac{1}{m}\log NFA\left(W,A_j,S\right) \tag{10.5}$$

When trying to rearrange the estimated formula, it is possible to edit the following:

$$Meaning\left(w,a_j\right)=-\frac{1}{m}\log\left(\begin{matrix}k\\m\end{matrix}\right)-[(m-1)\log N] \tag{10.6}$$

$$Score(w)=\max_{a_j\in A}\left(Rank\left(w,a_j\right)\right) \tag{10.7}$$

The meaning of a sentence in a set of j c is that a given word can be understood as gradually meaningful, important, or refined for that class. Phrases, with a very high average score, are associated with keywords that gradually illuminate the specific set. Again, it is expected that a class score is added to the brand selection and the selection of best R features. This grouping focuses on making the strategy. Placement factors are sorted using significance scores for each class. For example, the main component of each configured reorder denotes the word reference indicator. Now, part evaluation is used in each class in contrast to their centralized scores. When a summary of the solitaire components that depend on this class is documented, the largest position for all items is selected [6].

10.3.4 Cyberbully Detection

In cyberbullying, ML recognizes the incriminating words in both content and comments of a tweet. After receiving profitability at the preliminary processing stage, the class performance document is sent to order calculations. A prepared classifier is used here to perform the said actions. The training dataset contains a summary of computerized terms. With a preparatory dataset, a preprocessed Twitter dataset is tested with or without agony. An OGHOCNN search classifier is mainly used to identify the cyberbully words in real and tweet observations. This strategy equally distinguishes people who tweet a digitally imperious cyberbully. In this research, the OGHOCNN technique exhibited best output on high spatial data by paying more attention to the characteristics that surround the words. For the classification problem, the authors attempted to get more space in the text, given their short length and the tendency to focus on cyberbullying.

The grasshoppers are often found in nature making it one of the largest groups among all living things. The size of the ensemble makes the designers intimidating and dreamy. Grasshoppers are a group of insects that enjoy with a lot of fun as it is nymph and mature. As the tongue arrives, the cylinders of millions of grasshopper's nymphs are moved. These grasshoppers eat all the plants in its path. Subsequently, as they mature, they warm up in the air. The grasshoppers can migrate far and wide. In its larval phase, the grasshoppers move in a slow fashion while the small phases are the main characteristics of the herd. In old age, rapid movements and long reach are the main features of the screen. The main feature of grasshoppers harvesting is its search for a food source. The nature-inspired algorithms that logically divides the search process into two is explained in Section 10.1, one being exploratory and the other being exploitative. In the study section, the research agents are encouraged to go quickly as they become local at the time of exploitation. The target search and the associated two actions are performed by grasshoppers. By mathematically discovering this behavioral model, it is possible to design a new way of being inspired by nature.

Step 1: The mathematical model used to imitate the behavior of a group of grasshoppers is as follows:

$$Q_i = T_i + G_i + V_i \tag{10.8}$$

where Q_i is the location of an ith grasshopper, T_i is the social interaction, G_i is the gravity on ith grasshoppers, and V_i is the wind progress.

Step 2: To change the traditional grasshopper algorithm, an opposite method was introduced. Based on the opposition based learning (OBL) introduced by the current agent and their opposing agent, both are considered at the same time in order to get a better approximation for the solution of the current agent. The counter-agent solution is believed to be closer to the optimal global solution when compared to a random agent solution. The positions of the blocks of opposite dispersion (Oq_m) are entirely determined by the components q_m.

$$Oq_m = \left[oq_m^1, oq_m^2, \ldots, oq_m^d\right] \tag{10.9}$$

where $Oq_m = Low_m + Uq_m - q_m$ with $Oq_m \in [Low_m, Uq_m]$ is the position of the mth block with little scatter Oq_m in the dth size of the opposite blocks.

Step 3: To ensure random behavior, the equation can be written as follows: $q_i = r_1\ S_i + r_2\ F_i + r_3 W_i$, where r_1, r_2, and r_3 are random numbers in [0, 1].

$$T_i = \sum_{\substack{j=1 \\ j \neq i}}^{N} t\left(d_{ij}\right)\hat{d}_{ij} \tag{10.10}$$

where d_{ij} is the distance between the ith and jth grasshoppers, calculated as $d_{ij} = |q_j - q_i|$, s is a function that determines the strength of social forces $\hat{d}_{ij} = \frac{q_j - q_i}{d_{ij}}$: it is a uniform vector from the ith grasshopper to the jth grasshopper.

Step 4: The process of s determining social forces is considered as follows:

$$S(r) = Ae^{\frac{-r}{\int}} - e^{-r} \tag{10.11}$$

where W denotes the intensity of the attraction and $\int$ is the length of the attraction. The work should show how this affects the social interaction (attraction and affection) of grasshoppers.

Step 5: The form of the function s in this interval and component F of equation (10.5) are calculated as follows:

$$G_i = h\hat{e}_h \tag{10.12}$$

where g is gravitationally stable and $\hat{e}_g$ shows a unit vector toward the center of the earth.

Step 6: The V factor in equation (10.1) is determined as follows:

$$V_i = u\hat{e}_v \tag{10.13}$$

where c is a constant float and $\hat{e}_v$ is a direction of wind of unit vector.

Step 7: Nymph grasshoppers have no wings. So their development is closely related to the direction of the wind. If T, G, and V are taken in equation (10.14), this condition can be created as follows:

$$Q_i = \sum_{\substack{j=1 \\ j\neq i}}^{N} s\left(|q_j - q_i|\right)\frac{q_j - q_i}{d_{ij}} - h\hat{e}_g + c\hat{e}_v \tag{10.14}$$

where $t, T(r) = Ae^{\frac{-r}{l}} - e^{-r}$, and N are the number of locusts that sprite grasshoppers fall to the ground and their position should not be below a certain limit. Consequently, this is also not used in this research for reimplementation and progress measurement, as it prevents the calculation from examining the request space around a response. Most of the space is empty. Equation (10.15) shows that the communication between grasshoppers is valuable for imitating them:

$$Q_i = \left(\sum_{\substack{j=1 \\ j\neq i}}^{N} \frac{ub_d - lb_d}{2} s\left(|q_j^d - q_i^d|\right)\frac{q_j - q_i}{d_{ij}} \right) + \hat{T}_d \tag{10.15}$$

where ub_d is upper bound in the Dth dimension, and lb_d is possibly present in the Dth dimension $T(r) = Ae^{\frac{-r}{l}} - e^{-r}$, this is the measure of Dth dimension completion (the best adjustment so far) and the decreasing coefficient of u safe space, shock zone, and ambition. It is to be noted that T is practically the same part of equation (10.15) as T. However, the gravity is not considered (without the G segment), while the wind direction is accepted to be (A component) always aimed at the target $\hat{T}_d$. The following region of the locust is resolved based on its current position and is shown in equation (10.16). This capability considers only the grasshopper position and flow area, which corresponds to different grasshoppers. Here, the total number of grasshopper states to determine the range of search specialists around the target is considered.

$$B = b\max - \frac{b\max - b\min}{L} \tag{10.16}$$

Here c max is the maximum value, c min is the minimum value, l is the current cycle, and L is the maximum number stress. The best-case scenario is updated so far in each iteration. Also, the coefficient b is determined using equation (10.16). The separation between grasshoppers is standardized in each cycle [1, 5]. An element is updated periodically until the last rule is completed. The position and appropriateness of a good goal is the best essay for a global ideal. Only one layer of convention and maxpooling with three different channel sizes was used. Spans of three folding channels are selected as 1, 2 and 3, which is a specialty in terms of channels. The filter sizes were chosen based on the number of persistent words needed to understand the fear content. Here, m is a continuous word conversion process and its results are as given in equation (10.17):

$$F_i = f\left(w_c x_{i:i+m} + bc\right) \tag{10.17}$$

where $x_{i:i}+_m$, f_i, w_c, b_c, and f are the grid for adding m words, the estimation of the function-generating features, the weight of the relative folding filter, the counter-weight, and the staging work, respectively. A straight rectifier was used for implementation work. The most intense pool operation was applied to all capabilities from a single convolution channel. At this time, articles were consolidated into h, a vector of articles with sizes corresponding to the number of filters used. To determine the probability of Class m, a sensitive layer with the greatest failure at the pool exit was applied:

$$M\left(Y = i \mid X, \theta\right) = soft\max_i\left(w_s h + b_s\right) \tag{10.18}$$

Here X, f, Y, w_s, b_s, i, and θ are the information frameworks of installation, vectors of feature from crash and pool levels, class prediction, end-level loads, inclination to compare, and class number or parameters, respectively. The comparison framework is clear in the assessment area as shown below. So the cyberbullying is increased with a high degree of consistency.

10.3.5 Dataset Description

The dataset used in the search was obtained from Twitter using Twitter API. A total of 39,000 tweets were considered. Despite this, after recording the tweets, it was identified that there is an unequal class problem (there are not many threatening tweets). This is because the researchers inquire about the Twitter API without much attention, leading to return of scary tweets. As a result, the information was verified and deleted, evacuating copies and tweets with only images or URLs. A brief overview of the information collected for preparation and tests is presented in Table 10.3. The work was distributed and adequate guidelines were provided. For subjective purposes, a 25-question test was required to take the recognition of members into account. Finally, two people were selected among those who completed the assessment with 95% level.

10.4 RESULT AND DISCUSSION

The evaluation of the proposed calculation focuses on a trial study of the most significant realities. To begin with, OGHOCNN offered preferable results over conventional detection of cyberbullying. In addition, a content-based position was updated to evaluate various indicators such as failure and review for emotional evaluation. All tests were conducted using a Windows PC with 12 GB RAM and the execution was completed using MATLAB 2012.

10.4.1 Evaluation Metrics

Since recognition of cyberbullying is an ordering task, position accuracy is considered as the undisputed measurement solution. In any case, this is an unequal class problem. As a result, if accuracy is only considered as a measure, it is possible to achieve 80% accuracy, essentially identifying all test tweets with a large class of parts. This problem was understood by considering two different dimensions: visibility and precision. All measurements were recorded concurrently.

$$\text{Accuracy} = \frac{TP + TN}{TP + TN + FP + FN}$$

$$\text{Precision} = \frac{TP}{TP + FN}$$

$$\text{Recall} = \frac{TP}{TP + FP}$$

At this time, a complete correlation was made between the next three discoveries about cyberbullying. The goal is to show that the proposed calculation of OGHOCNN improves the current state of recognition of cyberbullying and it offers better expectations (greater accuracy), despite eliminating the need to create opportunities. The progress of the analysis begins with testing various qualities of channels, centers, pools and neurons to demonstrate that a change in the quality changes the nature of prediction. Table 10.1 shows the performance analysis of statistical measurements such as accuracy, precision, and recall in the proposed approach.

Figure 10.3 and Table 10.1 shows that the output of the proposed method, it is visible that the accuracy was 97%, precision was 98%, and recall was 97%. Figure 10.4 shows the comparison of proposed approach with existing accuracy measures, Figure 10.5 shows precision measure and Figure 10.6 shows recall measure of the proposed system.

TABLE 10.1 Performance Measures of the Proposed Research

No. of Neurons	Accuracy (%)	Precision (%)	Recall (%)
4	70	98	80
16	97	97	92
32	97	93	83
43	97	87	97
54	96	90	90

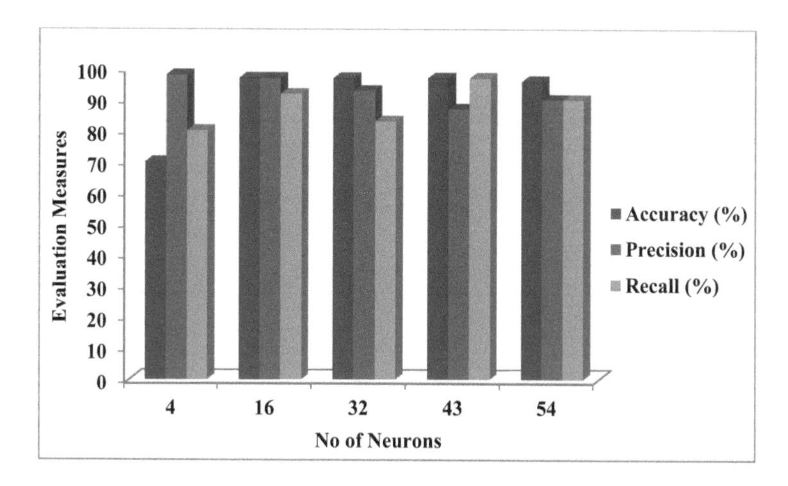

FIGURE 10.3 Graphical representation of proposed evaluation measures.

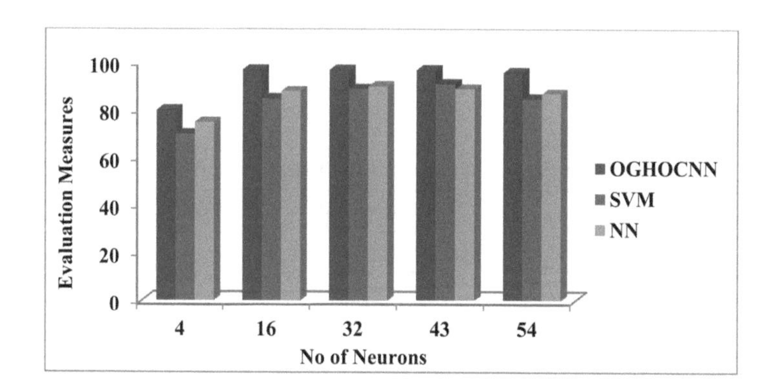

FIGURE 10.4 Graphical representation of the proposed and existing accuracy measures.

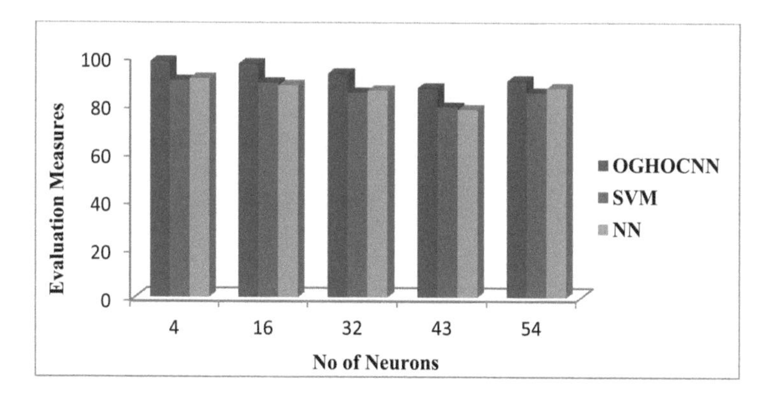

FIGURE 10.5 Graphical representation of proposed and existing precision measures.

TABLE 10.2 Comparison of the Proposed and Existing Accuracy Measures

No. of Neurons	OGHOCNN Accuracy (%)	SVM	NN
4	80	70	75
16	97	85	88
32	97	89	90
43	97	91	89
54	96	85	87

TABLE 10.3 Comparison of the Proposed and Existing Precision Measures

No. of Neurons	OGHOCNN Precision (%)	SVM	NN
4	98	90	91
16	97	89	88
32	93	85	86
43	87	79	78
54	90	85	87

TABLE 10.4 Comparison of Proposed and Existing Recall Measures

No. of Neurons	OGHOCNN Recall (%)	SVM	NN
4	80	75	77
16	92	85	87
32	83	77	76
43	97	89	85
54	90	85	88

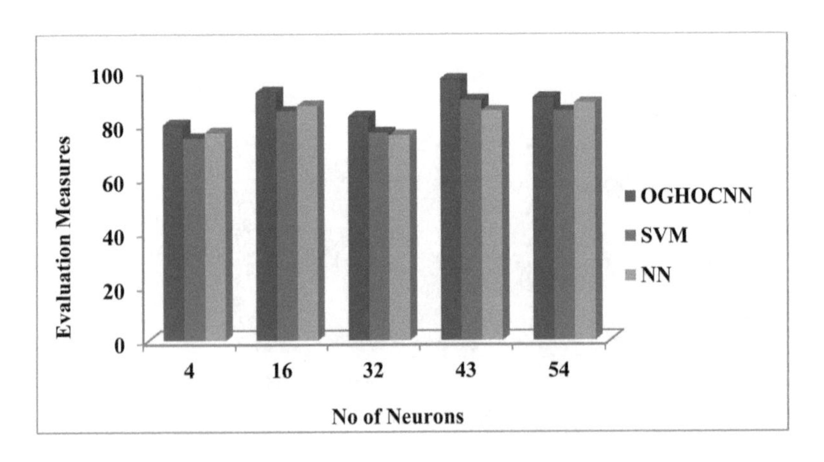

FIGURE 10.6 Graphical representation of proposed and existing recall measures.

10.4.2 Comparative Analysis

The OGHOCNN method suggests that the characteristic work of Twitter dataset is useful in comparing SVM and NN techniques. The proposed representation of Twitter tweets dataset was analyzed by modifying the classifier, and the recorded results are as follows:

The time taken by various classifiers to predict the detection of cyberbullying is shown in Tables 10.2–10.4. The experimental results are presented in connection with the proposed prediction for cyberbullying detection using comparative classification classifier, OGHOCNN. Machine Learning Ability and Selectivity OGHOCNN describe the best accuracy, precision and recall activity. The precise accuracy measures determine the appropriateness of cyberbullying detection technology. The refresh rate was accurate, while the prediction for classification of a cyberbullying detection-based ranking model (RM) classifier was highly reliable. The accuracy, precision, and recall percentages for FP and FN coefficients got improved.

10.5 CONCLUSION

This research work proposed the design and development of classification RM based on supervised feature selection to detect cyberbullying on Twitter. This is aimed at improving the accuracy of detection and reduction in the execution times. The selection of supervised functions was carried out using the classification strategy. OGHOCNN was used to perform detection. The dataset was collected from Twitter tweets. The OGHOCNN classifier was implemented in MATLAB while the results offered better precision.

REFERENCES

1. R. Pawar and R. R. Raje, "Multilingual Cyberbullying Detection System," *2019 IEEE International Conference on Electro Information Technology (EIT)*, Brookings, SD, 2019, pp. 40–44.
2. S. Parime and V. Suri, "Cyberbullying Detection and Prevention: Data Mining and Psychological Perspective," *2014 International Conference on Circuits, Power and Computing Technologies [ICCPCT-2014]*, Nagercoil, 2014, pp. 1541–1547.
3. A. Kovačević, "Cyberbullying Detection Using Web Content Mining," *2014 22nd Telecommunications Forum Telfor (TELFOR 2014)*, Belgrade, 2014, pp. 939–942.
4. H. Rosa, J. P. Carvalho, P. Calado, B. Martins, R. Ribeiro and L. Coheur, "Using Fuzzy Fingerprints for Cyberbullying Detection in Social Networks," *2018 IEEE International Conference on Fuzzy Systems (FUZZ-IEEE)*, Rio de Janeiro, 2018, pp. 1-7.
5. I. Ting, W. S. Liou, D. Liberona, S. Wang and G. M. Tarazona Bermudez, "Towards the Detection of Cyberbullying based on Social Network Mining Techniques," *2017 International Conference on Behavioral, Economic, Socio-Cultural Computing (BESC 2017)*, Krakow, 2017, pp. 1-2.
6. S. Tomkins, L. Getoor, Y. Chen and Y. Zhang, "A Socio-Linguistic Model for Cyberbullying Detection," *2018 IEEE/ACM International Conference on Advances in Social Networks Analysis and Mining (ASONAM 2018)*, Barcelona, 2018, pp. 53–60.
7. X. Zhang et al., "Cyberbullying Detection with a Pronunciation Based Convolutional Neural Network," *2016 15th IEEE International Conference on Machine Learning and Applications (ICMLA 2016)*, Anaheim, CA, 2016, pp. 740–745.
8. R. Zhao and K. Mao, "Cyberbullying Detection Based on Semantic-Enhanced Marginalized Denoising Auto-Encoder," *IEEE Transactions on Affective Computing*, vol. 8, no. 3, pp. 328–339, 1 July-Sept. 2017.

9. A. Upadhyay, A. Chaudhari, Arunesh, S. Ghale and S. S. Pawar, "Detection and Prevention Measures for Cyberbullying and Online Grooming," *2017 International Conference on Inventive Systems and Control (ICISC 2017)*, Coimbatore, 2017, pp. 1–4.
10. M. Yao, C. Chelmis and D. Zois, "Cyberbullying Detection on Instagram with Optimal Online Feature Selection," *2018 IEEE/ACM International Conference on Advances in Social Networks Analysis and Mining (ASONAM 2018)*, Barcelona, 2018, pp. 401–408.
11. D. Zois, A. Kapodistria, M. Yao and C. Chelmis, "Optimal Online Cyberbullying Detection," *2018 IEEE International Conference on Acoustics, Speech and Signal Processing (ICASSP 2018)*, Calgary, AB, 2018, pp. 2017–2021.
12. J. Zhang, T. Otomo, L. Li and S. Nakajima, "Cyberbullying Detection on Twitter Using Multiple Textual Features," *2019 IEEE 10th International Conference on Awareness Science and Technology (iCAST 2019)*, Morioka, Japan, 2019, pp. 1–6.
13. R. Sugandhi, A. Pande, S. Chawla, A. Agrawal and H. Bhagat, "Methods for Detection of Cyberbullying: A Survey," *2015 15th International Conference on Intelligent Systems Design and Applications (ISDA 2015)*, Marrakech, 2015, pp. 173–177.
14. Y. J. Foong and M. Oussalah, "Cyberbullying System Detection and Analysis," *2017 European Intelligence and Security Informatics Conference (EISIC 2017)*, Athens, 2017, pp. 40–46.
15. M. A. Al-Ajlan and M. Ykhlef, "Optimized Twitter Cyberbullying Detection Based on Deep Learning," *2018 21st Saudi Computer Society National Computer Conference (NCC 2018)*, Riyadh, 2018, pp. 1–5.
16. V. Banerjee, J. Telavane, P. Gaikwad and P. Vartak, "Detection of Cyberbullying Using Deep Neural Network," *2019 5th International Conference on Advanced Computing & Communication Systems (ICACCS 2019)*, Coimbatore, India, 2019, pp. 604–607.
17. T. Mahlangu, C. Tu and P. Owolawi, "A Review of Automated Detection Methods for Cyberbullying," *2018 International Conference on Intelligent and Innovative Computing Applications (ICONIC2018)*, PlaineMagnien, 2018, pp. 1–5.
18. D. Mouheb, M. H. Abushamleh, M. H. Abushamleh, Z. A. Aghbari and I. Kamel, "Real-Time Detection of Cyberbullying in Arabic Twitter Streams," *2019 10th IFIP International Conference on New Technologies, Mobility and Security (NTMS 2019)*, Canary Islands, Spain, 2019, pp. 1–5.
19. I. Nazar, D. Zois and M. Yao, "A Hierarchical Approach for Timely Cyberbullying Detection," *2019 IEEE Data Science Workshop (DSW 2019)*, Minneapolis, MN, 2019, pp. 190–195.
20. H. Nurrahmi and D. Nurjanah, "Indonesian Twitter Cyberbullying Detection Using Text Classification and User Credibility," *2018 International Conference on Information and Communications Technology (ICOIACT 2018)*, Yogyakarta, 2018, pp. 543–548.
21. A. Pascucci, V. Masucci and J. Monti, "Computational Stylometry and Machine Learning for Gender and Age Detection in Cyberbullying Texts," *2019 8th International Conference on Affective Computing and Intelligent Interaction Workshops and Demos (ACIIW 2019)*, Cambridge, UK, 2019, pp. 1–6.

CHAPTER 11

An Energy-Efficient Quasi-Oppositional Krill Herd Algorithm-Based Clustering Protocol for Internet of Things Sensor Networks

K.M. Baalamurugan,[1] R. Gopal,[2] D. Vinotha,[3] A. Daniel,[4] and Vijay Ramalingam[5]

[1]*School of Computing Science and Engineering, Galgotias University, Noida, Uttar Pradesh*

[2]*Information and Communication Engineering, College of Engineering, University of Buraimi, Al Buraimi, Oman*

[3]*Department of Computer Science & Engineering, Annamalai University, Tamil Nadu, India*

[4]*School of Computing Science and Engineering, Galgotias University, Greater Noida, Uttar Pradesh, India*

[5]*School of Computing Science and Engineering, Galgotias University, Uttar Pradesh, India*

CONTENTS

11.1 INTRODUCTION

Generally, Internet of Things (IoT) models are deployed actively and offer massive facilities for humans and business applications. By default, the IoT devices would be interlinked and communicates with one another by interchanging the accumulated sensor Information. Hence, the performance metrics like operating systems (OS), interacting protocols, and hardware units are applicable to provide enormous promising issues in developing IoT networks. IoT middleware environments and studies have been developed for resolving these issues [1]. For middleware platforms, the combined developing platform has been applied, and the response time from devices is limited by resource management objectives. Such middleware environments are mainly applied for minimizing the energy utilization, as the machines in IoT is often implemented on battery power. Hence, sensing tasks utilizes sensing period strategies. In absence of data variation, the devices are computed on the basis of predefined sensing period that consumes maximum energy by transmitting unwanted data resources.

The modern tools involved in IoT systems like sensors, actuators, power modules, tools, interaction methods, etc. The basic architecture of IoT sensor networks is shown in Figure 11.1. It is operated with the help of power that exists in incorporated battery. The operational features of devices are composed of massive effect on power application value. On the contrary, a wireless sensor network (WSN) has an exclusive base station (BS), where the devices are composed of an IoT network that interacts with one another with no limitation; hence, the drastic difference of IoT services could be offered to the users. In order to accomplish this, the devices should be capable of transmitting or interchanging the data with related devices. As the sensors, transmitter and actuator devices have been applied in this application, and then the energy utilization becomes more sensible and higher. Releasing maximum energy in devices is composed of a negative impact in prolonged IoT networks, and subsequently the earlier studies are concentrated on the effective way of utilizing energy, and examining the low power for sensing task, minimal energy communications, and extended battery application with no replacement of batteries or recharging the batteries [2].

When the energy is released completely, it results in expiry of a node and minimizes the network duration. In order to resolve these limitations, studies developed on storing node energy and extending the network duration that are considered to be the major issues in WSN [3]. In addition, clustering model could minimize the power application of nodes and elongate the networks existence [4, 5]. A WSN network is classified as various subsets and

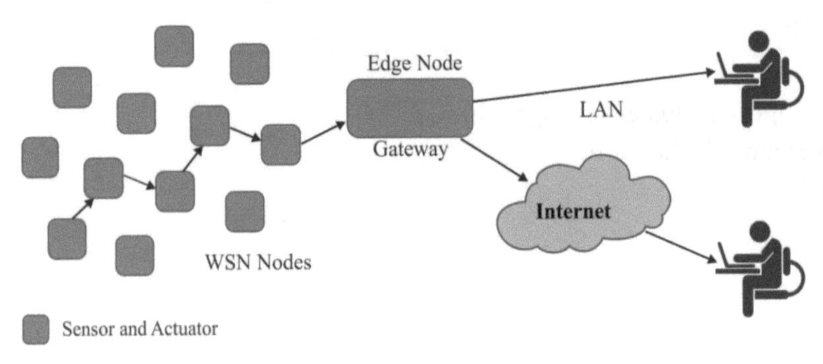

FIGURE 11.1 Architecture of IoT-sensor networks.

so called as clusters. The nodes present in identical cluster are adjacent to one another. For all clusters, a single node is named as cluster head (CH), and alternate nodes are termed as the cluster members (CMs). In spite of interacting with sink in a direct manner, members are applied for sending the sensed information to CHs where the collected data is transmitted to sink node. When compared to earlier planar routing models, LEACH is applicable to offer maximized network lifetime [6, 7]. Therefore, LEACH decides the CHs on the basis of the value produced by a node and previous fixed threshold that develops massive clusters and CH election is carried out in a random manner. But LEACH cannot ensure the optimal clustering model from energy management of the whole network.

Using the development of a LEACH, massive WSN clustering technologies are presented. It is composed of various clustering models on the basis of various clustering metrics like network architecture, communication method, topology, and stable routing. Here, the clustering methods are divided into two classes: independent and cooperative, on the basis of BS that is involved in clustering process. Autonomous clustering models on network nodes with no sink's contribution, and control data has been interchanged over all the nodes. The predefined clustering models can extend the network lifetime; however, it suffers from limitations with respect to working function. The independent methods are implemented on all nodes, and massive control data should be interchanged which results in extra power application. Though the cooperative models are involved in limiting the power application of control data exchange, this fails to obtain optimal clustering from complete network as it is composed of several parameters that affect the clustering and such attributes conflict mutually. Specifically, using the deployment of nature-based methods, the clustering techniques have been applied for enhancing clustering function recently [8].

An effectual node ranking LEACH (NRLEACH) for WSN has been projected. Here, the nodes are classified as two phases like clustering setup and data transmission. The NR model has been applied for CH election. NR approach has also been applied for calculating the grade of all nodes by count of iterations, which depends upon different parameters. Energy-centered clustering-based routing technique is presented in Ref. [9], which classifies the complete network to various static grids, and the nodes present in all grids are comprised of a cluster. The node estimates the rank based on the residual energy (RE) and maximum distance to alternate nodes in a cluster. Initially, each node telecasts the rank value for other nodes similarly. The node with maximum rank value can be selected as a CH. A new fuzzy clustering model is developed in Ref. [10] for 3D WSN named as FCM-3 WSN. Here, the developers have introduced an arithmetic approach of clustering in 3D WSN that considers the network power application, communication metrics, and the features of 3D network. The Lagrange multiplier (LM) approach has also been applied for determining the cluster centres as well as the membership matrix used for clustering operation.

A novel energy-efficient centroid-based routing technique (EECRP) is presented for WSN. In EECRP, the application of "energy centroid" has been presented. In Ref. [11], Zhang et al. have provided an enhanced LEACH method for WSN for optimizing the CH selection. First, the BS telecasts the initialization message to the all systems. Once the initialization message is received, nodes determine the distance from corresponding BS on the basis of obtained signal strength and save the data of adjacent nodes in a table. In Ref. [8], a novel power-effective multihop routing method (MR) has been developed. In MR, nodes are classified into three classes: CHs, member nodes, and independent nodes (IN).

A new particle swarm optimization (PSO) relied energy-efficient CH selection model for WSN (PSO-ECHS) has been presented. In PSO-ECHS, the applied attributes assumed, like intracluster distance, distance from CHs to BS, and the impact of RE on clustering outcome, and a linear integration of such variables have been applied in fitness function (FF) of particles.

A PSO-based method for routing as well as clustering in WSN has been presented in Ref. [12]. In case of clustering technique, the power application of CHs and CM nodes are considered. While using PSO for clustering the WSN, the coding model of particles is the dimension of particles and similar to count of network nodes, and location component of all dimension of particles is meant to be arbitrary value r that follows uniform distribution from (0, 1). An integer k is attained from r using transformation formula, and k is declared as CH-ID where the nodes of particle position. Moreover, an FF is mainly applied for reducing the power application of data transmission and manages the RE in transmission path.

In Ref. [13], Kuila and Jana presented a power-efficient clustering as well as routing methodologies for WSN such as PSO model. Here, the particle encoding approach for clustering is similar, but the objective of FF is applied to build a CH with maximized network lifetime and CM has smaller average distance to CH. In Ref. [14], Wang et al. recommended an enhanced routing framework with certain clustering with the help of PSO approach for heterogeneous WSN (EC-PSO). Initially, CHs are selected on the basis of geographical position. If the network is implemented for a period and power of nodes is not the same, EC-PSO is utilized for clustering the WSN.

This chapter presents an efficient metaheuristic quasi-oppositional krill herd (QOKHC) based clustering algorithm for IoT sensor networks. The proposed QOKHC algorithm involves the selection of CHs and organizes clusters. The presented model integrates the quasi-oppositional based learning (QOBL) in the krill herd (KH) algorithm to increase the convergence rate. The efficiency of the QOKHC-based clustering technique has been assessed and the results are examined under diverse measures: RE, network lifetime, alive node analysis, and the number of packets transmitted to BS.

11.2 THE PROPOSED CLUSTERING ALGORITHM

The working process of the QOKHC algorithm is shown in Figure 11.2. As depicted in the figure, the IoT sensor network undergoes node deployment in a random manner. Once the nodes are deployed, the nodes are initialized and information exchange takes place. Then, the node executes the QOBL algorithm and selects the CHs in an appropriate way. After the CHs are properly chosen, the nearby nodes join the CHs and construct clusters. Finally, data transmission from CMs to CHs takes place and reaches the BS via intercluster communication.

KH is a new optimized method that is used to solve difficult problems. Krill swarms hunt for food and communicate through members of the swarm type the fundamental of these techniques. Three stages are executed that are frequent in these KH methods; an optimal solution is continued by the explored ways. It involves three tasks in that the location of a krill has been set:

- Attempt inclined by other krill
- Foraging performance
- Physical diffusion

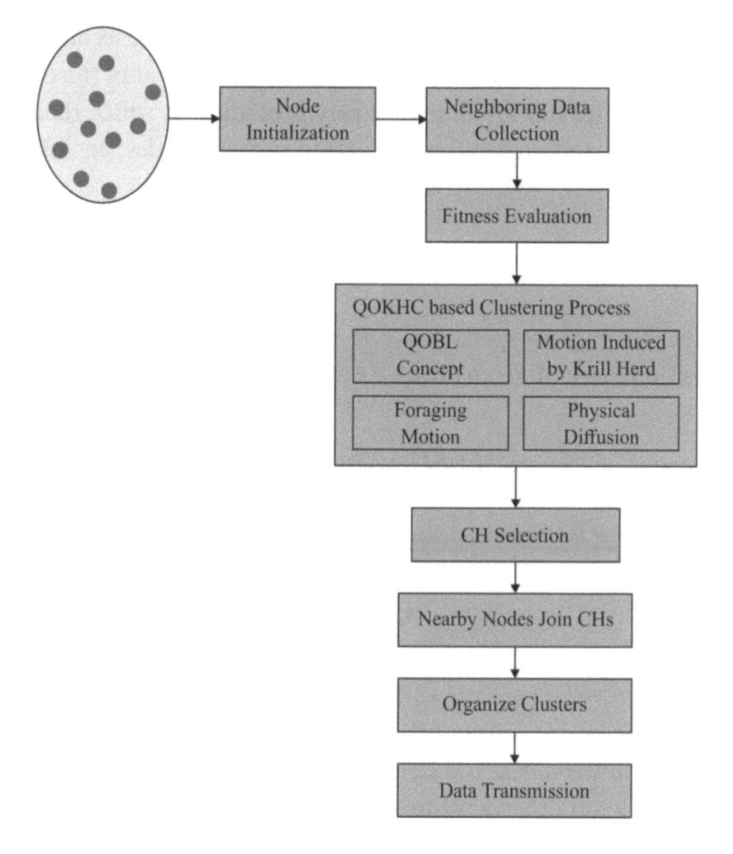

FIGURE 11.2 Block diagram of the QOKHC algorithm.

The KH considered the Lagrangian method, as illustrated in equation (11.1):

$$\frac{dX_i}{dt} = N_i + F_i + D_i \tag{11.1}$$

where N_i is the motion developed with other krill, F_i is the exploring motion, and D_i is the physical sharing; $i = 1, 2, \ldots, NP$ and NP is the size of the population. In the initial motion, the aim, local, and repulsive result are defined in its motion way α_j. For krill i, equation (11.2) is indicated by

$$N_i^{new} = N^{\max} \alpha_i + \omega_n N_i^{old} \tag{11.2}$$

where $N^{\max}$ signifies the maximal tempted speed, ω_n is the inertia weight, and N_i^{old} signifies the final motion. In second motion, the place of the food and past experience are denied. For the ith krill, it is determined in equations (11.3) and (11.4):

$$F_i = V_f \beta_j + \omega_{f F_i^{old}} \tag{11.3}$$

$$\beta_j = \beta_i^{food} + \beta_i^{best} \tag{11.4}$$

where V_f is the seeking speed, ω_f represents the inertia weight to the second motion, and F_i^{old} is the final motion; β_i^{food} is the food attractive, and β_i^{best} is the result of the optimal fitness of the ith krill so far. The third motion is an arbitrary procedure in which there are two parts: a highest diffusion speed and an arbitrary directing vector. Besides, equation (11.5) can be expressed by

$$D_i = D^{\max} \delta \tag{11.5}$$

where $D^{\max}$ signifies the highest speed flow and δ refers to an arbitrary vector. Using three processes, the position of a krill from t to $t+\Delta t$ can be signified, as given in equation (11.6):

$$X_i(t+\Delta t) = X_i(t) + \Delta t \frac{dX_i}{dt} \tag{11.6}$$

where Δt is attained from equation (11.7):

$$\Delta t = C_t \sum_{j=1}^{NV} \left(UB_j - LB_j\right) \tag{11.7}$$

where NV indicates the entire count of variables, LB_j and UB_j are likewise the lower and upper bounds of the jth variables, and C_t is the constant number 0.5. The progress of typical KH is influenced by another krill to suitable count of generations or until stopping condition is met, foraging and physical diffusion maintain for occurring. The flowchart of KH algorithm is depicted in Figure 11.3.

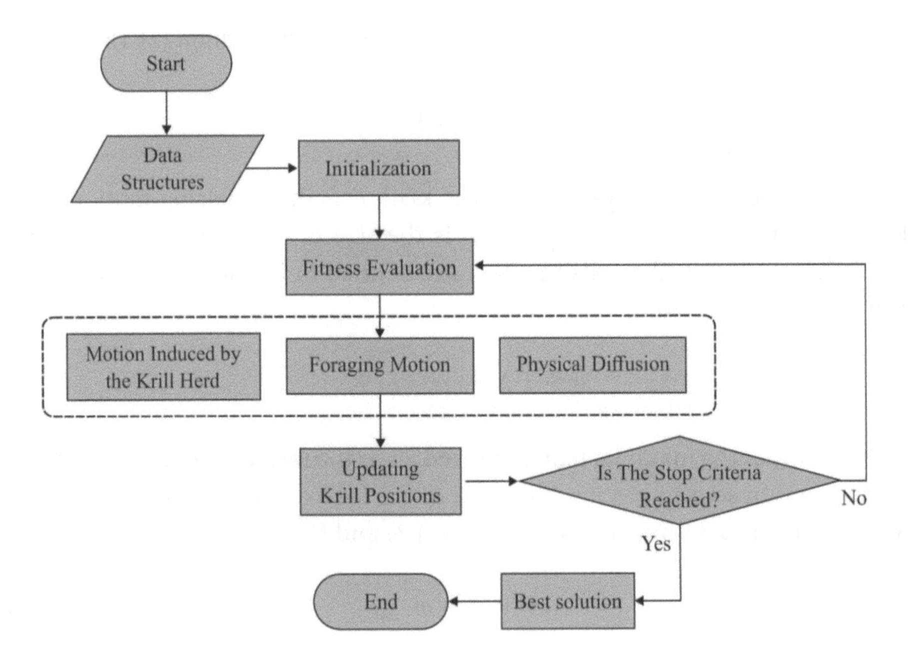

FIGURE 11.3 Flowchart of KH algorithm.

An OBL method is applied for accelerating the convergence and enhances the quality of solutions with regard to the present solutions and opposite solutions synchronously. At the beginning of the probability theory, the arbitrary solution is 50% superior to the opposite solutions and vice versa. So, the higher solution among the two inverse solutions is selected as the candidate solution that is improved explores the efficiency of evolutionary techniques. The OBL model has been efficiently executed to a different type of issues. In OBL, the models of opposite number as well as opposite point are determined in the following:

Opposite number: When x is an arbitrary number in the explore region $[a, b]$, its opposite number is illustrated as

$$x^o = a + b - x \tag{11.8}$$

Opposite point: When $P(x_1, x_2, \ldots, x_i, \ldots, x_d)$ is a point in d-dimensional space where $x_i \in [a_i, b_i]$, its opposite point $OP(x_1^o, x_2^o, \ldots, x_i^o, \ldots, x_d^o)$ can be determined as follows:

$$x_i^o = a_i + b_i - x_i ; i = 1, 2, \ldots, d \tag{11.9}$$

But it might be pointed out that the OBL has few enhancement methods, in that QOBL has been implemented by several researchers and verified to be efficient compared to OBL. Furthermore, the quasi-opposite number and the quasi-opposite point are determined as follows.

Quasi-opposite number: A quasi-opposite number x^{qo} of an arbitrary number x in the explore region $[a, b]$ is illustrated as follows:

$$x^{qo} = rand\left[\left(\frac{a+b}{2}\right), (a+b-x)\right] \tag{11.10}$$

Quasi-opposite point: A quasi-opposite point $QOP\left(x_1^{qo}, x_2^{qo}, \ldots, x_i^{qo}, \ldots, x_d^{qo}\right)$ in d-dimensional space is computed as follows:

$$x_i^{qo} = rand\left[\left(\frac{a_i + b_i}{2}\right), \left(a_i + b_i - x_i\right)\right]; i = 1, 2, \ldots, d \tag{11.11}$$

The QOBL is implemented not only in the initialize method but also in the evolutionary method of CS technique to update the population. In this chapter, the result is created by mutation method utilizing a quasi-opposite solution.

The communication situations of WSN take place in two ways: intercluster communication and intracluster communication. A broadcast in both varieties can be with a single hop to the BS or with multiple hops, initially to the CH and thereafter from CH to BS. Improving the intracluster communication and choosing a suitable cluster illustrative of every node in every round is the purpose of clustering. The data from many member nodes is combined at the CH and thereafter send to the BS. It can reduce the energy utilized. But a problem with these techniques is that the CH is constantly a suitable node that regularly loses its energy in the procedure. Thus, a node has to be allocated as the CH in all rounds. These decisions of choosing a fixed node are accepted by the KH. The fresh CH in a definite round is selected depending upon the energy taken by the node and distance from the member nodes that are not CH.

It mainly comprises four stages involving the clustering protocol function: (i) selecting the CH, (ii) constructing of clusters, (iii) collection of information, and (iv) communicating information. A set-up phase and the steady-state phase are the two phases. At first, in a single set-up stage, the sensor transmits the location and the remaining energy data to the BS. Then, the BS evaluates the average energy depending upon this information. In some provided round, a CH is chosen depending upon the maximum average energy in that specific round. So, the capable node is chosen as the CH to that round. A BS afterwards executes the technique to define the K count of fittest CHs. It diminishes the cost function, as given in equations (11.12)–(11.14):

$$\cos t = \beta \times f_1 + (1+\beta) \times f_2 \tag{11.12}$$

$$f_1 = \max_{k=1,2,\ldots,k} \left\{ \sum_{\forall n_i \in C_{p,k}} d\left(n_i, CH_{p,k}\right) / \left| C_{p,k} \right| \right\} \tag{11.13}$$

$$f_2 = \sum_{i=1}^{N} E(n_i) / \sum_{k=1}^{k} E\left(CH_{p,k}\right) \tag{11.14}$$

where f_1 represents the highest of average Euclidean distance of nodes to their connected CHs and $c_{p,k}$ is the count of nodes suitable to cluster C_k of krills. The function f_2 is determined as the ratio of entire first energy of each node $n_i, i = 1,2,\ldots,N$ in the network to the entire present energy of the CH candidates in the present round. β is a user-specified constant utilized for weight giving to all of the sub-objectives.

The purpose of the FF has been separated for minimizing the intracluster distance among the nodes as well as CHs simultaneously. It can be quantified by f_1; f_2 quantifies the energy efficiency in the network that is quantified. Based on the classification of the cost function above, a lesser value of f_1 and f_2 implies that the cluster contains an optimal count of nodes and has requisite energy to carry out the functions compared to a *CH*.

i. Allocate S krills to maintain K arbitrarily assigned to CHs among the fixed CH candidates

ii. Initialize krills using QOBL concept

iii. Compute the cost function of all krills
For all nodes, $n_i = 1,2,\ldots,N$.
Calculate distance $d(n_j, CH_{p,k})$ among node n_j and every $CH_{p,k}$.

Allocate node n_j to $CH_{p,k}$, where

$$d\left(n_i, CH_{p,k}\right) = \min\left\{d\left(n_i, CH_{p,k}\right)\right\} \tag{11.15}$$

Estimation of the cost function:

iv. Determine the optimal to all krills and determine the optimal positioned krills.

v. Update the positions separately in the explore space utilizing equations (11.16) and (11.17):

$$dXi = delta_t * \left(N(i) + F(i) + D(i)\right) \tag{11.16}$$

$$X(i) = X(i) + dXi \tag{11.17}$$

vi. Go to steps 2–4 until highest count of iteration is reached.

11.3 PERFORMANCE VALIDATION

The analysis of the experimental results of the QOKHC technique took place on diverse aspects. The results are examined in terms of energy, lifetime, and number of packets sent to BS.

Figure 11.4 depicts the network lifetime analysis of the QOKHC algorithm in terms of number of alive nodes under varying rounds. The figure stated that the alive nodes are high by the QOKHC algorithm, whereas alive nodes are minimum for the gray wolf optimization (GWO) algorithm. At the same time, the KH algorithm has tried to exhibit better network lifetime by attaining higher number of alive nodes over GWO algorithm, but failed to surpass QOKHC algorithm. In the execution round of 500, the number of alive nodes by QOKHC algorithm is 143 nodes, whereas minimal number of 88 and 112 alive nodes are exhibited by the GWO and KH algorithms, respectively.

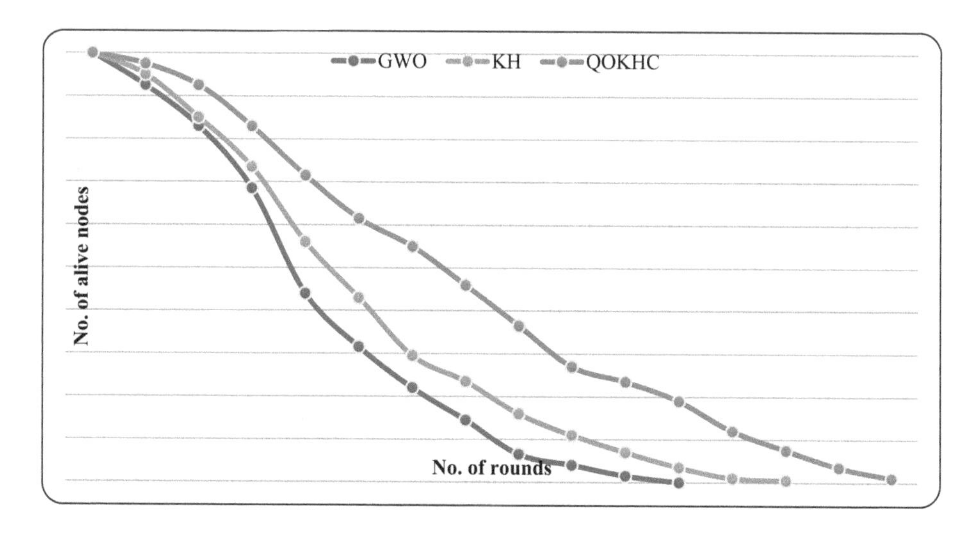

FIGURE 11.4 Alive node analysis of proposed QOKHC algorithm.

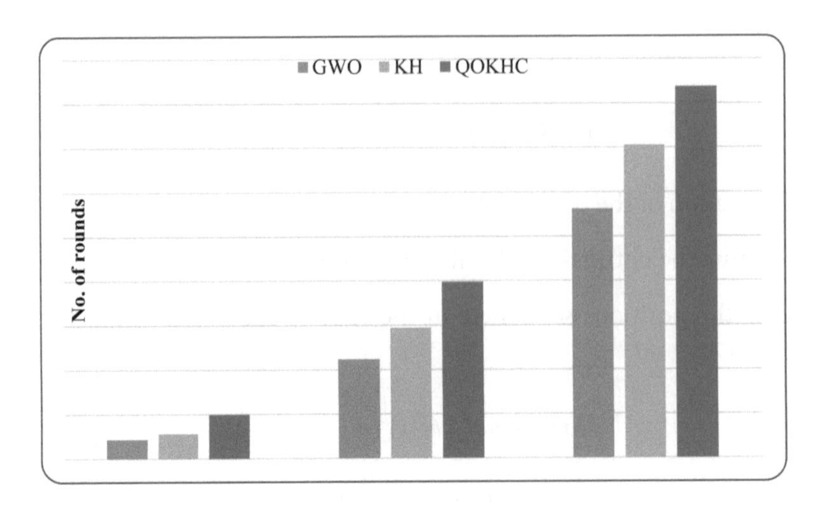

FIGURE 11.5 Network lifetime analysis of proposed QOKHC algorithm.

Similarly, under the execution round of 1,000, the number of alive nodes attained by the QOKHC algorithm is 54 nodes, whereas the GWO and KH algorithms have achieved minimal alive node count of 8 and 22 nodes, respectively. Likewise, on the round number of 1,400, all the nodes become dead by the GWO algorithm and only one node stays alive by the KH algorithm. But a maximum of 15 nodes are alive by the QOKHC algorithm. These values ensured the betterment in the network lifetime of the QOKHC algorithm.

Figure 11.5 shows the network lifetime analysis of the QOKHC algorithm in terms of FND, HND, and LND. The figure portrayed that the QOKHC algorithm has delayed the FND to 198 rounds, whereas it occurred earlier in the round numbers of 87 and 112 rounds by the GWO and KH algorithms, respectively. At the same time, the HND occurs at the round number of 795 by the QOKHC algorithm, whereas the GWO and KH algorithms exhibit minimal lifetime with 447 and 588 rounds, respectively. Finally, all the nodes become dead at the rounds of 1,124 and 1,411 rounds by the GWO and KH algorithms, respectively. However, the QOKHC algorithm demonstrated maximum network lifetime by attaining LND at 1,675 rounds.

Figure 11.6 examines the analysis of the QOKHC algorithm in terms of the number of packets reaching the BS under varying rounds. The figure stated that the number of packets reaching the BS is high by the QOKHC algorithm, whereas packet count reaching the BS is minimum by the GWO algorithm. At the same time, the KH algorithm has tried to exhibit a higher number of packet count reaching the BS over GWO algorithm, but failed to surpass QOKHC algorithm. In the execution round of 500, the packet count reaching the BS by QOKHC algorithm is 2,880 nodes, whereas the minimal number of 1,598 and 2,350 packets reached the BS by the GWO and KH algorithms, respectively. Similarly, under the execution round of 1,000, the number of packets received by the QOKHC algorithm is 4,980 nodes, whereas by the GWO and KH algorithms the minimal packet counts reaching the BS are 2,760 and 4,230 nodes, respectively. Likewise,

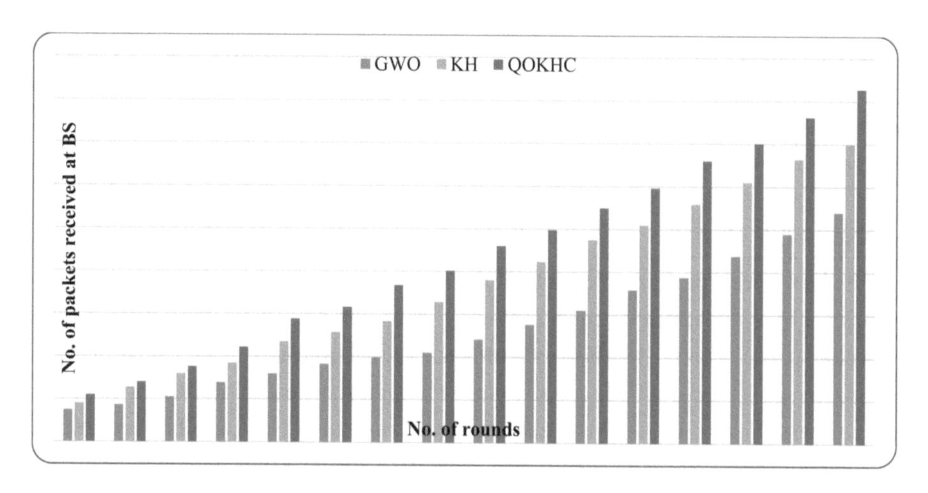

FIGURE 11.6 Analysis of number of packets reaching the BS by QOKHC algorithm.

in the round number of 1,400, maximum packet count reached the BS by the QOKHC algorithm. Therefore, the proposed method exhibits active network operation and more number of packets reaching the BS.

Figure 11.7 shows the average energy consumption analysis of the QOKHC algorithm and existing models. The figure portrayed that the QOKHC algorithm has demonstrated maximum residual energy compared to existing models. Besides, the residual energy is lower for the GWO and KH algorithms compared to the QOKHC algorithm. In the execution round of 500, the average energy consumed by QOKHC algorithm is 0.34 J, whereas the higher energy consumption are 0.4 and 0.38 J, respectively. At the same time, in the execution round of 1,000, average energy consumed by the QOKHC algorithm is

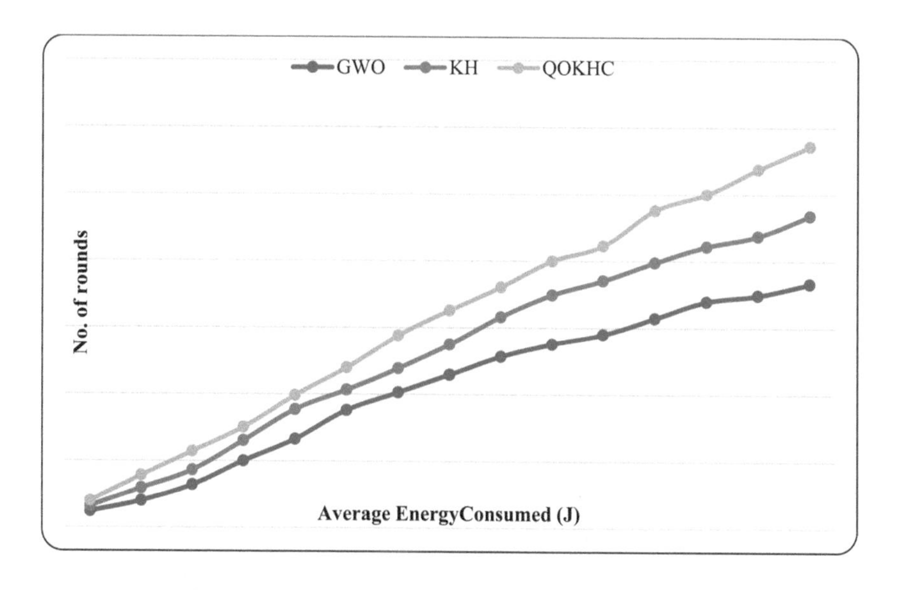

FIGURE 11.7 Energy efficiency analysis of the QOKHC method.

0.68 J, whereas the GWO and KH algorithms have achieved higher energy consumption of 0.75 and 0.62 J, respectively. These values portrayed that the QOKHC algorithm has consumed least amount of energy and offered maximum network lifetime over the compared methods.

11.4 CONCLUSION

This chapter has developed an efficient QOKHC-based clustering algorithm for IoT sensor networks. Once the nodes are deployed, the nodes are initialized and information exchange takes place. Then, the node executes the QOBL algorithm and selects the CHs in an appropriate way. After the CHs are properly chosen, the nearby nodes join the CHs and construct clusters. The analysis of the experimental results of the QOKHC technique takes place on diverse aspects and the results are examined under several aspects. The obtained simulation outcome depicted the proposed model in terms of network lifetime, energy, and number of packets sent to BS. As a part of future work, the performance of the QOKHC model can be improvised by the use of routing techniques.

REFERENCES

1. S. Li, L.D. Xu, and S. Zhao, "The Internet of Things: a survey," *Information Systems Frontiers*, vol. 17, pp. 243–259, 2015.
2. S. Arjunan and S. Pothula, "A survey on unequal clustering protocols in wireless sensor networks," *Journal of King Saud University: Computer and Information Sciences*, vol. 31, no. 3, pp. 304–317, 2019.
3. S. Arjunan and P. Sujatha, "Lifetime maximization of wireless sensor network using fuzzy based unequal clustering and ACO based routing hybrid protocol," *Applied Intelligence*, vol. 48, no. 8, pp. 2229–2246, 2018.
4. S. Arjunan, S. Pothula, and D. Ponnurangam, "F5N-based unequal clustering protocol (F5NUCP) for wireless sensor networks," *International Journal of Communication Systems*, vol. 31, no. 17, p. e3811, 2018.
5. J. Uthayakumar, T. Vengattaraman, and P. Dhavachelvan, "A new lossless neighborhood indexing sequence (NIS) algorithm for data compression in wireless sensor networks," *Ad Hoc Networks*, vol. 83, pp. 149–157, 2019.
6. N. A. Pantazis, S. A. Nikolidakis, and D. D. Vergados, "Energy-efficient routing protocols in wireless sensor networks: a survey," *IEEE Communications Surveys & Tutorials*, vol. 15, no. 2, pp. 551–591, 2012.
7. R. Singh and A. K. Verma, "Energy efficient cross layer based adaptive threshold routing protocol for WSN," *AEU – International Journal of Electronics and Communications*, vol. 72, pp. 166–173, 2017.
8. E. Alnawafa and I. Marghescu, "New energy efficient multihop routing techniques for wireless sensor networks: static and dynamic techniques," *Sensors*, vol. 18, no. 6, p. 1863, 2018.
9. A. Al-Baz and A. El-Sayed, "A new algorithm for cluster head selection in LEACH protocol for wireless sensor networks," *International Journal of Communication Systems*, vol. 31, no. 1, e3407, 2018.
10. D. T. Hai, L. H. Son, and T. L. Vinh, "Novel fuzzy clustering scheme for 3D wireless sensor networks," *Applied Soft Computing*, vol. 54, pp. 141–149, 2017.
11. Y. Zhang, P. Li, and L. Mao, "Research on improved low energy adaptive clustering hierarchy protocol in wireless sensor networks," *Journal of Shanghai Jiaotong University: Science*, vol. 23, no. 5, pp. 613–619, 2018.

12. M. Azharuddin and P. K. Jana, "PSO-based approach for energy-efficient and energy-balanced routing and clustering in wireless sensor networks," *Soft Computing*, vol. 21, no. 22, pp. 6825–6839, 2017.
13. P. Kuila and P. K. Jana, "Energy efficient clustering and routing algorithms for wireless sensor networks: particle swarm optimization approach," *Engineering Applications of Artificial Intelligence*, vol. 33, pp. 127–140, 2014.
14. J. Wang, Y. Gao, W. Liu, A. Sangaiah, and H.-J. Kim, "An improved routing schema with special clustering using PSO algorithm for heterogeneous wireless sensor network," *Sensors*, vol. 19, no. 3, p. 671, 2019.

CHAPTER 12

An Effective Social Internet of Things (SIoT) Model for Malicious Node Detection in Wireless Sensor Networks

Vijay Ramalingam,[1] Dinesh Babu Mariappan,[2] R. Gopal,[3] and K.M. Baalamurugan[4]

[1]*School of Computing Science and Engineering, Galgotias University, Uttar Pradesh, India*

[2]*Department of Computer Science and Engineering, Galgotia College of Engineering and Technology, Uttar Pradesh, India*

[3]*Information and Communication Engineering, College of Engineering, University of Buraimi, Al Buraimi, Oman*

[4]*School of Computing Science and Engineering, Galgotias University, Noida, Uttar Pradesh*

CONTENTS

12.1 INTRODUCTION

Internet of Things (IoT) has turned into a prevalent framework to help numerous advanced applications and services, for example, brilliant homes, smart healthcare, open security, modern observing, and condition assurance [1] to reinforce the unwavering quality and security in remote sensor systems. Hence, it is critical to structure a compelling security system for recognizing vindictive hubs in an IoT network [2]. IoT service enables certain capacities to be helped out through a predefined interface. A few researchers are especially keen on recognizing danger issues emerging during finding and coordinating information inside IoT [3]. The SIoT is a bigger social network, associating people and people and items, and articles and items. With numerous security issues in a SIoT model, it experiences a similar security problem as customary Internet-based as well as remote frameworks, including sticking, spoofing, and so forth [4]. The SIoT model is an interdisciplinary developing space that empowers self-ruling association among informal communication and the IoT. The nature of SIoTs presents different difficulties in its plan, design, execution, and activity of the executives [5]. SIoT has the ability to offer novel applications and networking services for IoT effectively [6].

Also, with the scope of social network service being extended from individual focused to a partnership focused, the association with IoT empowers a business practical joint effort [7]. The attackers change the conduct of the hubs in the system to fall and debase the usefulness of the wireless sensor networks. The malicious hubs in the wireless sensor systems can be identified utilizing cross-hybrid acknowledge scheme (HAS). In this strategy, the hubs in the wireless sensor system are gathered into a number of clusters. Each cluster ought to have just three hubs and an individual cluster key in all hubs in the cluster [8]. On the off chance that malicious hubs effectively alter the information, it can make an impact on the IoT work, that is, prompting an off-wrong decision [9].

A node is in the long run viewed as malicious when its total outcome progresses toward becoming lower than a specific limit. The disadvantage of this method is that it doesn't identify replay assaults distinguished by optimization, machine learning, and deep learning procedures [10]. From these strategies, the researchers have derived system insights and malicious node practices and announced the effective discovery of particular sending, Hello flooding, and sticking assaults by their interruptions [11]. In light of these insights and practices, the malicious nodes are effectively and adequately confined to SIoT arrangement.

12.2 REVIEW OF RECENT KINDS OF LITERATURE

In 2019, Ande et al. [12] proposed the next-generation Internet, a web system that fuses human qualities. Security dangers inside all design layers and some mitigation methodologies are discussed with some future advancements. Given the potentially sensitive nature of IoT datasets, there is a need to build up a standard for the sharing of IoT datasets among the examination and professional networks and other important partners. The potential of blockchain technology is encouraging in secure sharing of IoT datasets Banerjee et al. [13]. A methodology, deep learning, to empower the discovery of assaults in social web of things has been presented by Diro et al. [14]. The deep learning model is thought about against the conventional machine learning approach, and the appropriated assault discovery is assessed against the brought-together detection system. The tests have demonstrated that our disseminated assault discovery system is better than the combined detection systems utilizing deep learning model.

The design is to arrange the assaults that don't expressly harm the system; yet by contaminating the inside nodes, they are prepared to do the assaults on the system, which are named as inward assaults by Hajiheidari et al. [15]. At that point, categorizations of the IDSs in the IoT assault like denial of administration assault, Sybil assault, replay assault, particular sending assault, wormhole assault, black hole attack, sinkhole assault, sticking assault, false information assault have likewise been given utilizing regular highlights. The favorable circumstances and weaknesses of the chosen systems are also talked about. Parameter infusion, as a typical and incredible assault, is regularly abused by attackers to break into the servers of IoT by infusing malicious codes into the parameters of the solicitations, as explained by Yong et al. [16]. A hidden Markov model (HMM) based discovery framework is presented in this chapter, which is structured as a novel biregistry scoring design using both favorable and malevolent web traffic to protect against parameter infusion assaults in IoT systems.

12.3 NETWORK MODEL: SIoT

IoT is a developing worldview that aims to create the planet more astute in support of humankind [17–20]. Brilliant structure, smart health keen industry, savvy grid, keen home, shrewd transportation, and keen learning are a few constituents of a smart city. Social IoT based strategy aims for better living. SIoT condition encompassing an individual or association supports getting assortment and a lot of data, and gathered data gets the opportunity to be a clever service through the procedure of advanced obstruction [21]. For IoT conditions, the traffic elements and baselines are altogether different from heritage computer networks and consequently should be estimated and considered for IoT assault detection [22]. The SIoT network structure can be shaped as expected to guarantee efficient safety among the parts dependent on a few social perspectives with the goal of the discovery of administration, and this chapter successfully guides in arrangement synthesis for complex errands [23]. Second, versatility is guaranteed in informal communities through coordinated efforts [24–28]. The aim is to incorporate edge registering with group to keep up trust and protection runs in SIoT shown in Figure 12.1. This is not constantly substantial in certain conditions, which may influence the detection accuracy and even reason amassing faults.

Each of these issues is ruining the improvement of IoT. So to lessen the absolute IoT cost and offer data, we have to coordinate various capacities and assets into a bigger system [29–32]. This SIoT model has a few constraints or significant terms, which are clarified in the following [33–36]. These terms are significant to identify the malicious network model in the SIoT framework. The associated gadgets and their data are sensibly secure from

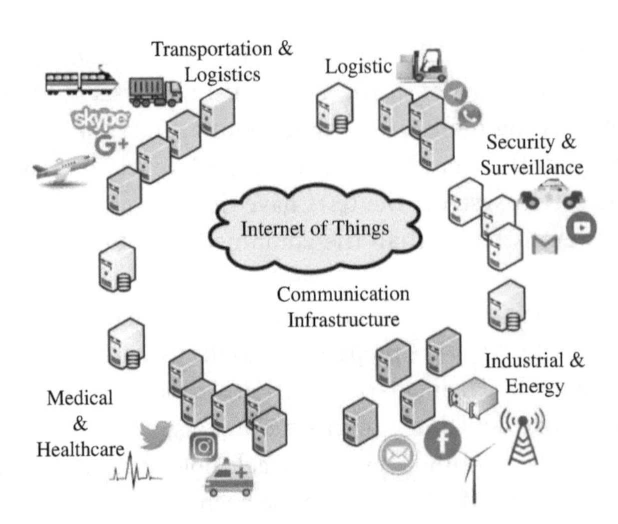

FIGURE 12.1 Graphical representation of SIoT.

abuse or damage, and the subsequent erosion of trust results in hesitance utilizing the Internet [37–40].

- A general IoT architecture is conveyed to assume control over open IoT applications with open working capacity [41–45].
- The service layer moderates approaching information solicitations and activation endeavors, permitting only approved associations through to the private cloud in SIoT [46–49].
- To locate the malicious nodes in the system model, picking an edge falsely to recognize malicious hubs from all hubs may lessen the detection accuracy [50].
- Since sensor hubs cannot know their neighbor's hubs before arrangement, there ought to be a neighbor's revelation period after the initial deployment and after any redeployment, empowering every sensor node to discover its neighbors [51].
- A little gathering of mobile malicious hubs can disturb these tasks in a single region at once as they move around in the system, making steering dark gaps in the SIoT model [52].

12.3.1 Malicious Attacker Model in SIoT

A general attacker model characterizes attacker activities and potential targets. In any case, past works accept that the attacker consistently arrives at its objective legitimately, which is false. The stronger malicious hubs that can dispatch a few assaults at the same time, for example, alter assault, drop assault, and replay assault. The attacker make use of portable malicious hubs with remote equipments in place of the actual hubs. An attacker can utilize radios with high signal solidarity to accomplish a comparable impact on moving hubs. The attacker benefits significantly from utilizing portable malicious

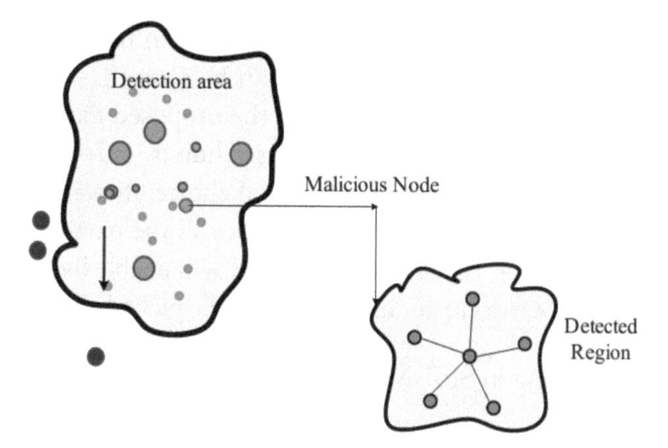

FIGURE 12.2 Detection of malicious node.

hub assaults, because of the expanded decent variety and the trouble of neighborhood identification and the attack model, shown in Figure 12.2.

An attacker can all the more explicitly dispatch a DDoS attack against the base station by having a lot of malicious hubs move to as many various areas as could reasonably be expected and flood the base station from every area turn. The attacker can likewise utilize a lot of mobile malicious hubs to disturb the different self-sorted out conventions, for example, directing, cluster development, time synchronization, and limitation. For example, mobile malicious hubs could produce and dispose of control bundles utilized in these conventions and in this manner make these protocols malfunction.

12.4 PROPOSED MN IN SIoT SYSTEM

Our proposed SIoT systems' malicious node recognition is significant; here exponential layer-based methodology is used for detection process. Information data is transmitted to the exponential kernel work to improve its learning procedure and acquire the last yield to the trust estimations of a node. To analyze the MN, at first the main cluster head (MCH) chosen from the SIoT is arranged by imaginative clustering procedures. In order to reduce energy consumption, the non-MCH nodes transmit the accumulated videos to the MCH. For secure directing, we have displayed trust-based Quality of Service (QoS) routing estimation. This computation removes the pernicious centers a long way from the gathering. This is on the grounds that picking a limit falsely to recognize malicious hubs from all hubs may decrease the detection accuracy. They at that point identified that the provenance may grow excessively quick with the expansion of bundled transmission jumps if system suspend. Choice of EK for MN detection assumes a critical role since it helps with mapping dataset to a higher dimensional space to acquire a superior understanding of the characterization model; this detection procedure is numerically represented in streaming equation:

$$\{\text{Node}_i\} \rightarrow \left\{ \begin{array}{l} (\text{node}(\text{node}_i^{'} \neq \text{MN}_i)\text{with proabaility y factor MCH} \\ (\text{node}'(\text{node}_i^{'} = \text{MN}_i)\text{with proabaility } (1 - \text{MCH}) \end{array} \right\} \tag{12.1}$$

Focusing on this issue, formalize the connection between the reputation of directing ways and hubs, by seeing that a hub's notoriety can be formalized as a multiple linear regression issue. To discover the MD according to the proposed technique, the heaviness of every sensor hub in some procedure that is a sensor hub is undermined and as often as possible it sends its report conflicting with an official choice, its weight is probably going to be diminished. Along these lines, there is option to call the part hub that is dispensed by the MCH as a general cluster. In this way, the MCH will enable live broadcasting of the assemble information from extra cluster individuals to the picked chosen vice cluster head.

12.4.1 Trust-Based Grouping in SIoT Network

The trust measure in MD discovery clustering plan is more energy-proficient and secure, which were the huge issues in asset-constrained sensor arrangement. They have additionally displayed a priority mechanism alongside the trust measurements, which was increasingly practical. In the event that the source hub advances the bundle through the neighbor hub accurately, at that point the trust estimation of that hub upsurges. On the off chance that the hub basically drops the bundle, at that point the trust estimation of that hub decreases and that hub is perceived as a malicious hub. Trust proportion of the SIoT framework is determined by utilizing conditions (12.2) and (12.3). A hub allows sending the parcel to a hub and then the immediate trust, as shown in equations (12.2) and (12.3):

$$T_{direct} = {T_{a,b}(t)} \Big/ {U_{a,b}(t)} \tag{12.2}$$

$$T_{indirect} = \frac{1}{b}\sum_{k=1}^{b} T_{neighbours}(t) \tag{12.3}$$

From equations (12.2) and (12.3), $T_{a,b}(t)$and$U_{a,b}(t)$ depict the number of parcels sent in the acknowledged manner by the node at the time and the number of bundles beneficially settled by hub from hub at a time $T_{indirect}$ It signifies the indirect trust degree that neighbors of node contain in the node by the determined time. It represents the neighbors of the 1-hop neighboring nodes.. The calculation of the indirect trust degree essentially relies on the neighbors' recommendations. Evaluating the trust level of every node with the help of circuitous suggestions brings several advantages. Here the social trust is assessed by the number of successful and unsuccessful changes between the cluster head and the node.

$$Trust\ measure = \alpha_1 T_{direct} + \alpha_2 T_{indirect} \tag{12.4}$$

A clustering-based trust instrument is proposed that tends to the security issues in IoT. It finds the similitude of enthusiasm for each cluster through the gauge the trust and incentive ahead of time. From the trust esteem portable sensors or a smart mobile with detecting, this speaks to the cluster heads' obligation in IoT network. The individuals from IoT are classified as hubs and MCH. In each cluster, the MCH gets data from each IoT hub and advances to the next cluster head. In the removed interchanges, the cluster head looks for the assistance of residual cluster heads to transmit the information to the goal hub in SIoT framework. When the clusters are shaped, the MN is recognized by EK strategy.

12.4.2 Exponential Kernel Model for MN Detection

In order to guarantee a high evaluation of efficiency for our malicious hub discovery procedure in SIoT system machine learning, EK is significant. The supernodes to detect malicious hubs, which could eliminate with the false criticism issue, existed in the input depending upon trust esteems, and it is graphically presented in Figure 12.3. Every single clustered node is remarkably recognized and know their own land position, which can be gotten utilizing a situating framework, for example, the sensors [16]. The security of the data in the remote sensor hubs is generally significant. The assailants change the conduct of the nodes in a network to crumble and corrupt the usefulness of the wireless sensor systems.

Definition I

The proposed kernel is a capacity whose worth relies upon the good ways from the beginning or from some point. In machine learning, the radial basis work kernel, or exponential kernel capacity is used as a support to help vector machine.

$$K(f, f_i) = \exp\left(-\alpha \|f - f_i\|^2\right), \quad \chi > 0 \tag{12.5}$$

where α is the certain mapping embedded in the exponential kernel; this capacity presents a set of N basis capacities, one for every data point q, which take the form $\alpha(\|c - c'\|)$ from some non-linear function, which will be discussed shortly. Along these lines, the nth value function depends on the distance, usually taken to be Euclidean, between two clustered parts. The yield of the mapping is then taken to be a linear mix of the fundamental elements of malicious information in SIoT.

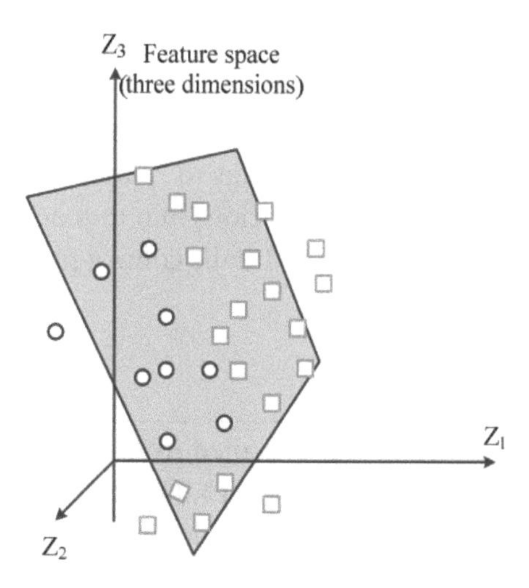

FIGURE 12.3 Trust values-based EK model.

12.4.2.1 Example of Proposed Detection System

- To analyze the probability of these assaults dependent solely on their harm or impact, in light of the fact that these assaults may influence one another.
- A malicious node could spread the false data that a standard node is suspicious and, with this single vote, viably dispose of from the system, possibly causing a swearing-of-service in huge pieces of the system.
- If there are in excess of two malicious nodes in a similar way, which perform two various assaults.
- The condition when it is difficult to analyze the probability of these assaults dependent exclusively on their harm or impact, in light of the fact that these assaults may influence one another.
- The passes a malicious node accept that the malicious hub can perform a SIoT arrangement.

12.4.3 Detection Model

It is characterized as the proportion between quantities of nodes effectively identified to all hubs. Detection stage is sorted into malicious hub detection ratio and non-malicious hub identification proportion. It is estimated in rate and it changes somewhere in the range of 0 and 100. The transmission is comprised of an assault and the data to locate the malicious node data.

$$Detection = \pm \text{Max}_{(Fixed\ value)}\left\{\beta + 10\log\left(\beta^2\right)\right\} \tag{12.6}$$

Relying upon the estimation of the value of its multiplier, a malicious transmission may or may not be detected as malicious. A few assaults are identified from our proposed system, which are as follows:

Worm hole attack (WHA): The wormhole attack is the larger part of serious security assaults that can significantly intrude on the interchanges over the system. Furthermore, it is tough to identify and simple to actualize. The attacker receives parcels at one area in the system, "tunnels" them to another area, and rematches them there in a wormhole attack.

Selfish node Attack (SNA): Selfish node helps to spare its own one of kind assets and takes extremely less control. This sort of malicious hub toss outs each of the bundles that it gets aside from those that are bound to it. It falls control parcels that consider the hubs would not be conceded in the directing.

12.5 RESULTS AND ANALYSIS

This proposed SIoT MN recognition was executed in NS2 with 300 nodes for recreation model. These nodes are performing in the district 1,000 m × 1,000 m with the transmission range 250 m. Reproduction parameters for recreation model are depicted in Table 12.1, in light of the leftover vitality of non-MCH and the execution results are investigated by confusion grid and some recognition rate to the comparison process.

TABLE 12.1 Simulation Parameters

Parameter	Value
Area size	1,000 m × 1,000 m
Number of nodes	300
Routing protocol	AODV
MAC	802_11
Antenna	Omni Antenna
Radio propagation model	Two-Ray Ground
Packet size	512 bytes
Initial transmitting power	0.660 W
Initial receiving power	0.395 W
Initial energy	10.3 J
Simulation time	100 s

Delivery ratio: It is the ratio of the measure of bundles got effectively and the total sum of parcels transmitted.

Throughput: It is the number of information that can be sent from the sources to the goal every second. Unit of this parameter is kb/s.

In the assessment, the detection perplexity lattice appears in Table 12.2; here two arrangements of data are referenced: the genuine and anticipated class of SIoT network system and the detection execution can be estimated utilizing both the accuracy rate and the error rate. This is on the grounds that the two models erroneously distinguished the low-positioning ordinary nodes as the malicious ones. At that point, Table 12.3 demonstrates the proposed consequences of MN detection in SIoT system: the measurements like throughput, delivery ratio, loss level, and proposed detection pace of the framework. For instance, the proposed throughput of 1.88 kb/s is thought to exist, that is in ANFIS and Bayesian methodology, the thing that matters is 5.52–10.12% compared to other relative regular strategies. Among the metrics, the significant factor is detection rate and data loss, these are discussed with the graphical portrayal. Abuse detection by identifying assaults by their realized marks doesn't require numerous assets; however, it has the disadvantage of not recognizing unpublished assaults. So to guarantee a high evaluation of effectiveness for our malicious hub detection procedure, a topology of SIoT has been provided. But this happens just when the objective hub and the one that is additionally tricked have nearly a similar separation to the made-up position of the malicious node just as to its genuine position. Toward the end, among the four methods of correlation, our proposed framework is better for MN detection on SIoT networks.

TABLE 12.2 Analysis of MN Detection Level

Confusion Matrix			
Malicious	TP	FN	Actual attacks
Benign	FP	TN	Malicious nodes
Results	Malicious nodes	Benign nodes	Actual benign

TABLE 12.3 Detection Level Analysis

Technique	Throughput (kb/bit)	Delivery Ratio (%)	Loss Ratio	Detection Ratio (%)
Proposed model	1.88	93.22	10.08	96.58
ANFIS	1.48	86.22	9.08	92.55
Bayesian approach	0.95	79.88	10.11	88.59
Trust model with a defense scheme	0.79	81.11	15.68	90.45
Epidemic routing	1.15	79.56	12.44	90.22

Figures 12.4 and 12.5 demonstrate the loss proportion and throughput of SIoT MN detection. At first the loss ratio is demonstrated in Figure 12.5 and in this investigation 300 nodes are selected for the MN detection. The analysis of proposed malicious node detection framework in SIoT concerning normal parcel loss ratio is presented. Loss ratio is low when there is less number of nodes and normal loss ratio is high when there is progressively the number of nodes. The information transmission is through SIoT system, transmission tally is utilized to consider interface quality. It guesstimates the number of retransmissions expected to push bundles by estimating the misfortune pace of communicating parcels in the midst of sets of neighboring nodes. On the off chance that the node size is expanded, the loss ratio to some degree is diminished, on the grounds that the execution time was huge. For instance, if the node is 175, the loss is 14.56%; in contrast, the products of 175 nodes is 11.29%; from this investigation, the area limited in proposed system, besides polynomial capacity, is utilized for ratio analysis. At that point, throughput of the system is shown in Figure 12.5, the exhibition of the proposed system is high if its throughput is high and the presentation of the proposed framework is low if its throughput is low. Here the throughput of MN detection approach in-stock model is investigated; here five cases are considered for the assessment. Best throughput on the off chance that five are the most extreme throughput worth is 1.99 kb/s for case 3. The above-mentioned graphs are used to demonstrate our SIoT arrangement system detection level as far as malicious hub detection process.

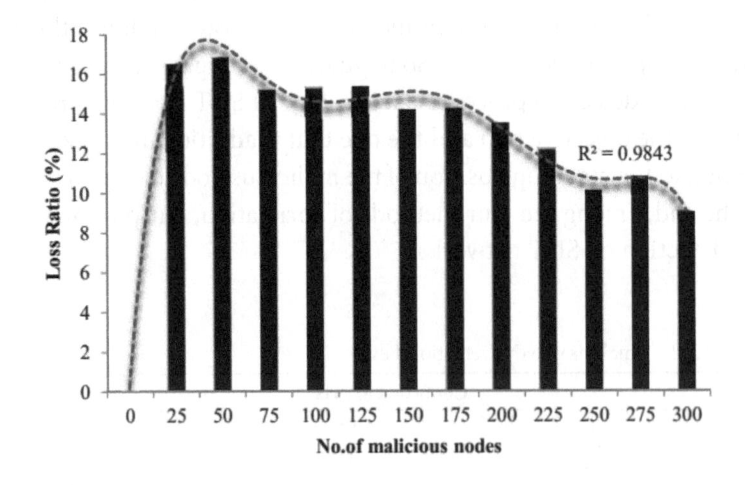

FIGURE 12.4 Loss ratio analysis for proposed model.

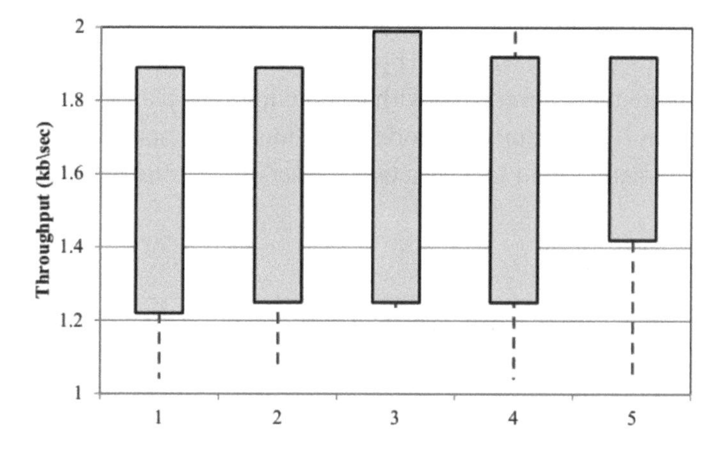

FIGURE 12.5 Throughput analysis.

Figure 12.6 shows MN detection in SIoT, the execution of the proposed malicious node detection system in SIoT condition is analyzed in terms of detection proportion. Test results demonstrate that our methodology can distinguish malicious nodes with stability and high accuracy, that is, improving the detection rate by 30–45% when contrasted with hard detection rate. It is estimated in rate and it changes between the levels of 0 and 300. The exhibition of the system is legitimately corresponding to the detection rate in both malicious and non-malicious cases. For instance, if node is 150, then detection proportion

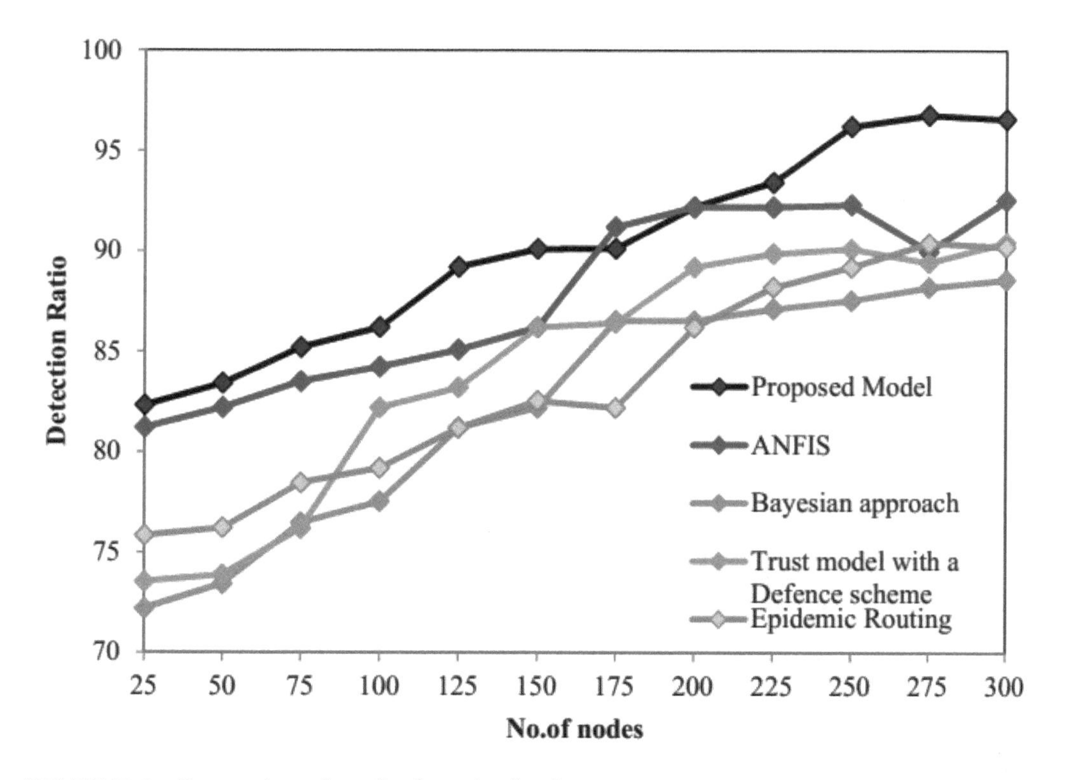

FIGURE 12.6 Comparison chart for detection level.

is 80–85% in Bayesian methodology and trust model. All ordinary procedures contrasted with the proposed model of EK with MCH process. The matchless quality of the proposed instrument is exhibited and contrasted with a contemporary plan for detection time and detection proportion. Our suspicious node detection relies upon trade of data among neighbors; this parameter ought to have a task to carry out in the detection pace of suspicious nodes.

12.6 CONCLUSION

One of the significant issues that are identified with the utilization of SIoT in cruel conditions is the hole in their security. Most existing algorithms ascertain the notoriety of all nodes dependent on the directing way or data transmission in SIoT arrangement. The recreation results demonstrate that the proposed model get the greatest detection level of SIoT network systems. The malicious nodes are recognized by getting the acknowledgments from the goal nodes and the exhibition of the proposed framework is analyzed utilizing parcel conveyance ratio and dormancy is significant. As the size of malicious nodes is small, the lifetime of sensor node is additionally extremely less and the battery life is likewise less. Our proposed model contrasted with some other existing regular strategies like ANFIS, trust model, etc.; from this comparison, most extreme conveyance proportion and detection rate is 9.08% and 96.58%, respectively, in SIoT detection model. The reproduction results showed that our model could successfully distinguish malicious conduct, for example, arrangement, Sybil, and record polluter. In future deep learning and some roused optimization mode will be utilized to distinguish the MN in SIoT network.

REFERENCES

1. A. Perrig, J. Stankovic, D. Wagner, "Security in wireless sensor networks", *Communications of the ACM*, Volume 47, Issue 6, Pages 53–57, 2004.
2. N. Gulati, P. D. Kaur, "Towards socially enabled Internet of Industrial Things: architecture, semantic model and relationship management", *Ad Hoc Networks*, Volume 91, Pages 101869, 2019.
3. R. Rohini, R. K. Gnanamurthy, "A simple and efficient malicious node detection system for improving the performance of the wireless sensor networks", *International Journal of Applied Engineering Research*, Volume 11, Issue 1, Pages 396–400, 2016.
4. A. M. Shabut, K. P. Dahal, S. K. Bista, I. U. Awan, "Recommendation based trust model with an effective defence scheme for MANETs", *IEEE Transactions on Mobile Computing*, Volume 14, Issue 10, Pages 2101–2115, 2014.
5. S. R. Sahoo, B. B. Gupta, "Hybrid approach for detection of malicious profiles in twitter", *Computers & Electrical Engineering*, Volume 76, Pages 65–81, 2019.
6. J. W. Ho, M. Wright, S. K. Das, "Distributed detection of mobile malicious node attacks in wireless sensor networks", *Ad Hoc Networks*, Volume 10, Issue 3, Pages 512–523, 2012.
7. E. M. Daly, M. Haahr, "Social network analysis for information flow in disconnected delay-tolerant MANETs", *IEEE Transactions on Mobile Computing*, Volume 8, Issue 5, Pages 606–621, 2008.
8. I. M. Atakli, H. Hu, Y. Chen, W. S. Ku, Z. Su, "Malicious node detection in wireless sensor networks using weighted trust evaluation," In *Proceedings of the 2008 Spring Simulation Multiconference* (pp. 836–843), April 2008. Society for Computer Simulation International.
9. D. I. Curiac, O. Banias, F. Dragan, C. Volosencu, O. Dranga, "Malicious node detection in wireless sensor networks using an autoregression technique," In *International Conference on Networking and Services* (ICNS'07) (pp. 83–83), June 2007, IEEE.

10. M. Y. Hsieh, Y. M. Huang, H. C. Chao, "Adaptive security design with malicious node detection in cluster-based sensor networks", *Computer Communications*, Volume 30, Issue 11–12, Pages 2385–2400, 2007.
11. M. Zabihimayvan, R. Sadeghi, H. N. Rude, D. Doran, "A soft computing approach for benign and malicious web robot detection", *Expert Systems with Applications*, Volume 87, Pages 129–140, 2017.
12. R. Ande, B. Adebisi, M. Hammoudeh, J. Saleem, "Internet of things: evolution and technologies from a security perspective", *Sustainable Cities and Society*, Volume 54, 2019.
13. M. Banerjee, J. Lee, K. K. R., Choo, "A blockchain future to Internet of Things security: a position paper", *Digital Communications and Networks*, Volume 4, Issue 3, 2017.
14. A. A. Diro, N., Chilamkurti, "Distributed attack detection scheme using deep learning approach for Internet of Things", *Future Generation Computer Systems*, Volume 82, Pages 761–768, 2018.
15. S. Hajiheidari, K. Wakil, M. Badri, N. J. Navimipour, "Intrusion detection systems in the Internet of Things: a comprehensive investigation", *Computer Networks*, Volume 160, Pages 165–191, 2019.
16. B. Yong, X. Liu, Q. Yu, L. Huang, Q. Zhou, Malicious web traffic detection for Internet of Things environments. *Computers & Electrical Engineering*, Volume 77, Pages 260–272, 2019.
17. M. Elhoseny, M. S. Mahmoud, K. Shankar, "Optimal deep learning based convolution neural network for digital forensics face sketch synthesis in Internet of Things (IoT)", *International Journal of Machine Learning and Cybernetics*, July 2020. Doi: https://doi.org/10.1007/s13042-020-01168-6.
18. M. Sivaram, E. L. Lydia, I. V. Pustokhina, D. A. Pustokhin, M. Elhoseny, G. Prasad Joshi, K. Shankar, "An optimal least square support vector machine based earnings prediction of blockchain financial products", *IEEE Access*, Volume 8, Pages 120321–120330, June 2020.
19. K. Shankar, Y. Zhang, Y. Liu, L. Wu, C.-H. Chen, "Hyperparameter tuning deep learning for diabetic retinopathy fundus image classification", *IEEE Access*, Volume 8, Pages 118164–118173, June 2020.
20. A. Rajagopal, A. Ramachandran, K. Shankar, M. Khari, S. Jha, Y. Lee, G. Prasad Joshi, "Fine-tuned residual network-based features with latent variable support vector machine-based optimal scene classification model for unmanned aerial vehicles", *IEEE Access*, Volume 8, Pages 118396–118404, June 2020.
21. I. V. Pustokhina, D. A. Pustokhin, D. Gupta, A. Khanna, K. Shankar, G. Nguyen, "An effective training scheme for deep neural network in edge computing enabled Internet of Medical Things (IoMT) systems", *IEEE Access*, Volume 8, Issue 1, Pages 107112–107123, December 2020.
22. V. Porkodi, A. R. Singh, A. R. WahabSait, K. Shankar, E. Yang, C. Seo, G. Prasad Joshi, "Resource provisioning for cyber-physical-social system in cloud-fog-edge computing using optimal flower pollination algorithm", *IEEE Access*, Volume 8, Issue 1, Pages 105311–105319, December 2020.
23. I. V. Pustokhina, D. A. Pustokhin, J. J. P. C. Rodrigues, D. Gupta, A. Khanna, K. Shankar, C. Seo, G. Prasad Joshi, "Automatic vehicle license plate recognition using optimal *k*-means with convolutional neural network for intelligent transportation systems", *IEEE Access*, Volume 8, Issue 1, Pages 92907–92917, December 2020.
24. B. Sujitha, V. S. Parvathy, E. L. Lydia, P. Rani, Z. Polkowski, K. Shankar, "Optimal deep learning based image compression technique for data transmission on Industrial Internet of Things applications", *Transactions on Emerging Telecommunications Technologies*, April 2020, Doi: https://doi.org/10.1002/ett.3976.
25. N. Krishnaraj, M. Elhoseny, E. L. Lydia, K. Shankar, O. ALDabbas, "An efficient radix trie-based semantic visual indexing model for large-scale image retrieval in cloud environment", *Software: Practice and Experience*, April 2020. Doi: https://doi.org/10.1002/spe.2834.

26. A. F. Saviour Devaraj, M. Elhoseny, S. Dhanasekaran, E. L. Lydia, K. Shankar, "Hybridization of firefly and improved multi-objective particle swarm optimization algorithm for energy efficient load balancing in cloud computing environments", *Journal of Parallel and Distributed Computing*, Volume 142, Pages 36–45, August 2020. Doi: https://doi.org/10.1016/j.jpdc.2020.03.022.
27. S. N. Mohanty, E. L. Lydia, M. Elhoseny, M. M. Gethami AlOtaibi, K. Shankar, "Deep learning with LSTM based distributed data mining model for energy efficient wireless sensor networks", *Physical Communication*, Volume 40, June 2020. Doi: https://doi.org/10.1016/j.phycom.2020.101097
28. J. Samuel Raj, S. Jeya Shobana, I. V. Pustokhina, D. A. Pustokhin, D. Gupta, K. Shankar, "Optimal feature selection based medical image classification using deep learning model in Internet of Medical Things", *IEEE Access*, Volume 8, Issue 1, Pages 58006–58017, December 2020.
29. K. Shankar, A. RahamanWahabSait, D. Gupta, S. K. Lakshmanaprabu, A. Khanna, H. M. Pandey, "Automated detection and classification of fundus diabetic retinopathy images using synergic deep learning model", *Pattern Recognition Letters*, Volume 133, Pages 210–216, May 2020. Doi: https://doi.org/10.1016/j.patrec.2020.02.026.
30. J. Uthayakumar, M. Elhoseny, K. Shankar, "Highly reliable and low-complexity image compression scheme using neighborhood correlation sequence algorithm in WSN", *IEEE Transactions on Reliability*, February 2020. Doi: https://doi.org/10.1109/TR.2020.2972567.
31. K. Geetha, V. Anitha, M. Elhoseny, K. Shankar, P. Shamsolmoali, M. M. Selim, "An evolutionary lion optimization algorithm-based image compression technique for biomedical applications", *Expert Systems*, January 2020. Doi: https://doi.org/10.1111/exsy.12508.
32. S. N. Mohanty, K. C. Ramya, S. S. Rani, D. Gupta, K. Shankar, S. K. Lakshmanaprabu, A. Khanna, "An efficient lightweight integrated block chain (ELIB) model for IoT security and privacy", *Future Generation Computer Systems*, Volume 102, Pages 1027–1037, January 2020.
33. S. Sankhwar, D. Gupta, K. C. Ramya, S. S. Rani, K. Shankar, S. K. Lakshmanaprabu, "Improved grey wolf optimization-based feature subset selection with fuzzy neural classifier for financial crisis prediction", *Soft Computing*, Volume 24, Pages 101–110, 2020.
34. S. Famila, A. Jawahar, A. Sariga, K. Shankar, "Improved artificial bee colony optimization based clustering algorithm for SMART sensor environments", *Peer-to-Peer Networking and Applications*, Volume 13, Pages 1071–1079, 2020.
35. K. Shankar, M. Elhoseny, "Trust based cluster head election of secure message transmission in MANET using multi secure protocol with TDES", *Journal for Universal Computer Science*, Volume 25, Issue 10, Pages 1221–1239, 2019.
36. A. K. Dutta, M. Elhoseny, V. Dahiya, K. Shankar, "An efficient hierarchical clustering protocol for multihop Internet of vehicles communication", *Transactions on Emerging Telecommunications Technologies*, Volume 31, Issue 5, May 2020.
37. M. Elhoseny, K. Shankar, J. Uthayakumar, "Intelligent diagnostic prediction and classification system for chronic kidney disease", *Scientific Reports*, Volume 9, Page 9583, 2019.
38. M. Elhoseny, K. Shankar, "Reliable data transmission model for mobile ad hoc network using signcryption technique", *IEEE Transactions on Reliability*, Volume 69, Issue 3, Pages 1–10, June 2019.
39. K. Shankar, S. K. Lakshmanaprabu, A. Khanna, S. Tanwar, J. J. P. C. Rodrigues, N. R. Roy, "Alzheimer detection using group grey wolf optimization based features with convolutional classifier", *Computers & Electrical Engineering*, Volume 77, Pages 230–243, July 2019.
40. M. Elhoseny, K. Shankar, "Optimal bilateral filter and convolutional neural network based denoising method of medical image measurements", *Measurement*, Volume 143, Pages 125–135, September 2019.
41. V. K. S. Ragavan, M. Elhoseny, K. Shankar, "An enhanced whale optimization algorithm for vehicular communication networks", *International Journal of Communication Systems*, April 2019. https://doi.org/10.1002/dac.3953

42. S. K. Lakshmanaprabu, K. Shankar, R. S. Sheeba, A. Enas, N. Arunkumar, G. Ramirez, J. Uthayakumar, "An effect of big data technology with ant colony optimization based routing in vehicular ad hoc networks: towards smart cities", *Journal of Cleaner Production*, Volume 217, Pages 584–593, April 2019.
43. S. K. Lakshmanaprabu, K. Shankar, M. Ilayaraja, A. W. Nasir, V. V. N. Chilamkurti, "Random forest for big data classification in the Internet of Things using optimal features", *International Journal of Machine Learning and Cybernetics*, Volume 10, Issue 10, Pages 2609–2618, October 2019.
44. K. Shankar, M. Elhoseny, R. S. Kumar, S. K. Lakshmanaprabu, X. Yuan, "Secret image sharing scheme with encrypted shadow images using optimal homomorphic encryption technique", *Journal of Ambient Intelligence and Humanized Computing*, Volume 11, Pages 1821–1833, December 2020.
45. K. Shankar, M. Elhoseny, E. Dhiravidachelvi, S. K. Lakshmanaprabu, W. Wu, "An efficient optimal key based chaos function for medical image security", *IEEE Access*, Volume 6, Issue 1, Pages 77145–77154, December 2018.
46. D. Gupta, A. Khanna, S. K. Lakshmanaprabu, K. Shankar, V. Furtado, J. J. P. C. Rodrigues, "Efficient artificial fish swarm based clustering approach on mobility aware energy-efficient for MANET", *Transactions on Emerging Telecommunications Technologies*, Volume 30, November 2018.
47. S. K. Lakshmanaprabu, S. N. Mohanty, K. Shankar, N. Arunkumar, G. Ramireze, "Optimal deep learning model for classification of lung cancer on CT images", *Future Generation Computer Systems*, Volume 92, Pages 374–382, March 2019.
48. M. Elhoseny, K. Shankar, S. K. Lakshmanaprabu, A. Maseleno, N. Arunkumar, "Hybrid optimization with cryptography encryption for medical image security in Internet of Things", *Neural Computing and Applications*, Volume 32, Pages 10979–10993, October 2018. https://doi.org/10.1007/s00521-018-3801-x
49. S. K. Lakshmanaprabu, K. Shankar, D. Gupta, A. Khanna, J. J. P. C. Rodrigues, P. R. Pinheiro, V. H. C. de Albuquerque, "Ranking analysis for online customer reviews of products using opinion mining with clustering", *Complexity*, Volume 2018, Pages 1–9, September 2018.
50. K. Karthikeyan, R. Sunder, K. Shankar, S. K. Lakshmanaprabu, V. Vijayakumar, M. Elhoseny, G. Manogaran, "Energy consumption analysis of virtual machine migration in cloud using hybrid swarm optimization (ABC–BA)", *Journal of Supercomputing*, Volume 76, Pages 3374–3390, September 2018. Doi: https://doi.org/10.1007/s11227-018-2583-3
51. K. Shankar, M. Elhoseny, S. K. Lakshmanaprabu, M. Ilayaraja, R. M. Vidhyavathi, M. Alkhambashi, "Optimal feature level fusion based ANFIS classifier for brain MRI image classification", *Concurrency and Computation: Practice and Experience*, August 2018. Doi: https://doi.org/10.1002/cpe.4887
52. S. K. Lakshmanaprabu, K. Shankar, A. Khanna, D. Gupta, J. J. P. C. Rodrigues, P. R. Pinheiro, V. H. C. de Albuquerque, "Effective features to classify big data using Social Internet of Things", *IEEE Access*, Volume 6, Pages 24196–24204, April 2018.

CHAPTER 13

IoT-Based Automated Skin Lesion Detection and Classification Using Gray Wolf Optimization with Deep Neural Network

A. Daniel,[1] S. Venkatraman,[2] Srinath Doss,[3] Bhanu Chander Balusa,[4] Andino Maseleno,[5] and K. Shankar[6]

[1]*School of Computing Science and Engineering, Galgotias University, Greater Noida, Uttar Pradesh, India*

[2]*School of Computer Science and Engineering, Vellore Institute of Technology, Chennai, Tamil Nadu, India*

[3]*Faculty of Computing, Botho University, Gaborone, Botswana*

[4]*SCOPE, Vellore Institute of Technology, Chennai, Tamil Nadu, India*

[5]*STMIK Pringsewu, Lampung, Indonesia*

[6]*Department of Computer Applications, Alagappa University, Karaikudi, Tamil Nadu, India*

CONTENTS

13.1 INTRODUCTION

Nowadays, the smart applications like smart homes are embedded with interior devices and home management models become familiar. It enhances the lifestyle of a human being; however, it also enhances energy consumption. Additional merits are that it has been applied extensively by enhancing the security as well as safety using motion sensors or cameras that are fixed to offer the alert message that eliminates the burglary issue. Additionally, there are some significant aspects that simplify the development of modern homes. These events activate the effective prediction of diseases and maximize the health state of an individual. In the last decades, the cancer disease and its risk have maximized [1], and the skin disease has become a malignant one, named as melanoma. This mainly occurs in the pigmented nerve of the skin. If there is any modification in skin, then it refers to a mutation of genes and invocation of melanoma, named as malignant tumor. According to the survey, it has been stated that the melanoma exists in the maximal rate in human skin cancers, which has increased exponentially in recent years [2]. However, the earlier prediction of a disease helps in enhancing the patient's lifetime to the greater extent. When the disease prediction is performed, then it is hardly difficult to save the patient's life. In addition, the likelihood of metastasis is higher, where it results in minimal survival rate.

In order to create awareness regarding the disease, the health condition and the symptoms that are recognizable have to be managed and prevented. The patient's life can be saved by rapid prediction and offering better treatment. The best solution has to be provided without any limitations where the modern solutions are capable of examining the symptoms in the alternate events. Smart homes are composed of embedded sensors that are able to maximize the management system so that the survival rate can be improved by visiting a nearby clinic. The knowledge of people about the risk of cancer can be gathered from the sources like TV, Internet, newspapers, education system, communication exchange, etc. Globally, the developers have tried to develop solutions that can be applied practically with low cost and computational energy. The advanced studies in Internet of Things (IoT) are divided, which is one of the higher objectives in protecting the health state, which depends upon the people's behavior as well as sentiment.

Skin cancer is mainly caused because of the unlimited development of unwanted cells, which is highly dangerous and spreads all over the human body. It is a deadliest disease that leads to death in absence of earlier prediction. In general, skin cancer has been categorized as two cases: malignant melanoma and benign melanoma. Initially, melanoma is one of the fatal infections of cancer with higher morbidity globally [3]. Based on the survey of American Cancer Society (ACS), around 80,110 novel cases of melanoma have been analyzed and 9,830 peoples are dead in the United States in recent years. Additionally, 132,000 melanoma and 3 million normal cases were filed around the world. If the disease is detected in stage-I, then 5 years of lifetime can be improved, whereas it is reduced to 5%, if detected at stage-IV. As melanoma is highly infectious and reduces the lifetime of a human, it has acquired massive concentration from medical experts and development team. Hence, the major objective of a dermatologist is to detect the melanoma at previous stage. Therefore, it can be highly complex because of the distinct features of melanocytic and non-melanocytic skin lesions. However, it is difficult for a well-trained dermatologist to examine accurately.

In medical examination, dermatologists apply the dermoscopy for detecting the presence of melanoma. Dermoscopy is defined as the non-invasive image examination for pigmented skin lesions in dermatology [4]. The dermoscopy is performed by placing a gel on the affected skin area and using a magnifier, the infected region becomes more visible than seeing through the human eye. Such pigmented model as well as structures is applied for examining the melanoma under the application of diverse diagnosing systems like ABCD rule, 7-point checklist, and Menzies approach. Therefore, it is based on human vision, general training, and medical experience of dermatologist. As a result, dermoscopy can achieve maximum accuracy in melanoma diagnosis [5]. Therefore, in order to maximize the efficiency of melanoma analysis, programmed dermoscopic image investigation is applied [6]. Computer-based skin lesion prediction as well as classification is composed of various steps like preprocessing, segmentation, feature extraction, and classification. Some of the typical disadvantages in computer-aided detection (CAD) models are lesion area segmentation as well as continuous features selection (FS) from original data. Furthermore, it uses various methods of imaging modalities; hence, it suffers from demerits that make the process highly tedious.

Prediction of skin diseases while processing an image is meant to be the significant aspect for sensors and sensing technologies, especially for computational intelligence as well as image processing. Besides, massive studies have been developed on clinical symptoms of skin diseases and dedicated skin markers for chemical examination. In Ref. [7], it was projected how plasmacytoid dendritic cells, inflammatory dendritic epidermal cells, and Langerhans cells affect anti-viral skin defense approach. Schmidt and Zillikens [8] defined the diagnostic gold standard for skin diseases prediction on the basis of analyzing auto-antibodies as well as mucous membranes under the application of immunofluorescence microscopy and commercialized devices. Developers have projected the way of how antigens make changes in skin characterization for serological analysis. In Ref. [9], it explained a chemical approach for detecting atopic dermatitis, psoriasis, and contact dermatitis, which is simplified using transcriptomic profiling. Some of the threatening disease has been predicted with the help of protein expression levels. In Ref. [10], the skin lesions classification with the application of deep convolutional neural network (DCNN) has been presented. The newly developed approach undergoes training by applying pixels and disease labels for input. Hence, the studies have depicted that classification is carried out in two cases: keratinocyte carcinomas versus seborrhoeic keratosis, and malignant melanomas versus nevi.

In Ref. [11], the approach for melanoma skin cancer prediction is presented. Researchers implied a model that integrates deep learning (DL) along with presented skin lesions ensemble model. In Ref. [12], the optoacoustic dermoscopy approach has been applied on the basis of excitation energy as well as penetration depth measures for skin analysis using ultra-imaging. It has been represented for analyzing the absorption spectra at several wavelengths for visualization of morphological as well as functional skin features. The model for detecting skin lesion and melanoma depends upon DL method presented in Ref. [13]. The researchers have defined segmentation as well as feature extraction to the center of melanoma using lesion indexing network that applies fully convolutional residual network (FCRN), which is major prediction technology. In Ref. [14], it is identified with periodical review over diverse machine learning (ML) approaches used in skin diagnosis as well as diseases investigation.

Chao et al. [15] have made an extensive comparison of mobile applications for skin observation as well as melanoma investigation. Many studies are developed by comparing the apps, and explained the capability of effective image processing from android phone camera and sensing methods, which is provided with legal aspects of ethical, quality, and visible deployment of apps for the purpose of treatment. Apart from the research on image analysis, the establishment of sensing approach is transparency. In Ref. [16], a smart sensing method for observing human skin as well as pathogen microbiota has been deployed. It is referred to as a composition of nanowire and thin film metal oxide models. The machine applies GC-MS (gas chromatography-mass spectrometry) and SPME (solid-phase microextraction) as data sampling technologies. In Ref. [17], a review of hyperspectral imaging has been applied.

This study develops an automated skin lesion detection and classification model using gray wolf optimization with deep neural network model called GWO-DNN model. The GWO-DNN model involves image segmentation, feature extraction, and classification. Initially, the fuzzy c-means (FCM) based image segmentation is applied to identify the diseased portions in the dermoscopic image. Afterwards, the gray level co-occurrence matrix (GLCM) technique is employed for extracting the useful set of features. Finally,

GWO-DNN algorithm is employed as a classification technique for classifying the distinct class labels of the skin lesion images. The performance validation of the GWO-DNN model undergoes evaluation against ISIC dataset.

13.2 THE PROPOSED GWO-DNN MODEL

The working process of the GWO-DNN model is illustrated in Figure 13.1. As depicted, the input images undergo image preprocessing, segmentation, feature extraction, and classification. These processes are discussed in the subsequent sections.

13.2.1 FCM-Based Segmentation

In this study, FCM-based segmentation model is applied to identify the diseased areas of the skin lesion image. In general, the clustering method of objective function for fuzzy clustering is based on the scheme of separation of fuzzy groups. To enhance the method, the subsequent objective function is utilized:

$$J_2(u, v) = \sum_{i=1}^{c}\sum_{k=1}^{N}(u_{ik})^2(d_{ik})^2 \tag{13.1}$$

where c is the count of clusters, d_{ik} is the distance, N is the count of pixels of some gray image, u_{ik} is the connection degree of the kth pixel in the ith cluster, which fulfilled the situation $\Sigma_{i=1}^{c} u_{ik} = 1, \forall_{u_{ik}} \in [0,1]$. An extremely utilized clustering method is FCM. The objective

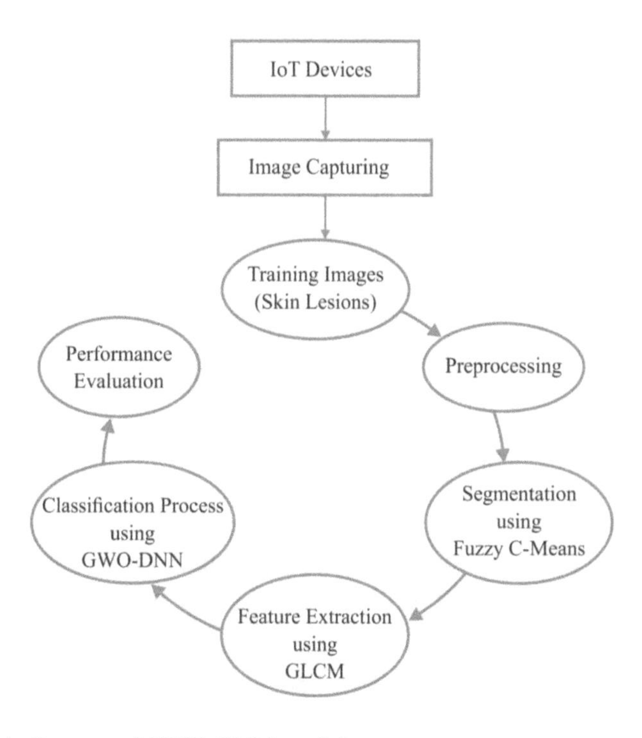

FIGURE 13.1 Block diagram of GWO-DNN model.

function aims minimizing with the iteration to attain the best image segmentation. An objective function of the usual FCM method is determined as follows:

$$J_m = \sum_{i=1}^{c}\sum_{k=1}^{N} u_{ik}^m \left\| x_k - v_i \right\|^2 \tag{13.2}$$

where $\{x_k, k = 1,2,\cdots, N\}$ is a dataset of pixels of some gray image, $\{v_i, i = 1,2,\ldots,c\}$ is the group of the cluster centers, and m is the index of the fuzzy weight. A novel value of relationship degree u_{ik} and the novel cluster centers v_i are computed utilizing the subsequent equation:

$$u_{ik} = \frac{\left(\left\| x_k - v_i \right\|^2\right)^{-1/(m-1)}}{\sum_{j=1}^{c}\left(\left\| x_k - v_j \right\|^2\right)^{-1/(m-1)}}, \; v_i = \frac{\sum_{k=1}^{N} u_{ik}^m x_k}{\sum_{k=1}^{N} u_{ik}^m} \tag{13.3}$$

But the method is effectively a local search method and it depends upon gradient descent. It also has a superior dependency on the first form.

13.2.2 Feature Extraction

The GLCM-based feature extraction is the model of extracting the features. But the efficiency of the provided GLCM-based features and the position of the texture features depend upon the gray level count employed. Some of the GLCM features involved in the proposed method are as follows.

13.2.2.1 Angular Second Moment (ASM)

ASM is said to be the evaluation of image homogeneity. A homogeneous image must be composed of a minimum number of gray levels as applied in GLCM; however, maximum values of $P(a,b)$ are utilized. Therefore, sum of squares is enhanced as given in the following:

$$ASM = \sum_{a=0}^{G-1}\sum_{b=0}^{G-1}\{P(a,b)\}^2 \tag{13.4}$$

13.2.2.2 Contrast

It means the value of local difference in an image. The contrast balances the $P(a,b)$ distant from diagonal $a = b$. When the image variations are maximum, then $P[a,b]$'s focus on higher diagonal, and enhance the contrast measure, which is described as follows:

$$Contrast = \sum_{n=0}^{G-1} n^2 \left\{\sum_{a=1}^{G}\sum_{b=1}^{G} P(a,b)\right\}, |a-b| = n \tag{13.5}$$

13.2.2.3 Inverse Difference Moment (IDM)

IDM was affected by image homogeneity. Since the weighting factor is $(1+(a-b)^2)^{-1}$, the IDM obtains minimum contributions from dissimilar regions (a,b). The simulation outcome accomplished may be lower IDM value for various images and higher value for similar images, as defined in equation (13.6):

$$IDM = \sum_{a=0}^{G-1}\sum_{b=0}^{G-1}\frac{1}{1+(a-b)^2}P(a,b) \tag{13.6}$$

13.2.2.4 Entropy

It defines the unstructured organization of an image. Difficult textures result in maximum entropy. It is highly remarkable and inversely connected to energy.

$$Entropy = -\sum_{a=0}^{G-1}\sum_{b=0}^{G-1}P(a,b)\times \log\left(P(a,b)\right) \tag{13.7}$$

13.2.2.5 Correlation

It describes the calculations of gray-level linear dependence from pixels at specific location from which the relationship is independent. Correlation may be higher when the image contains a manageable number of linear structure, as provided in equation (13.8):

$$Correlation = \sum_{a=0}^{G-1}\sum_{b=0}^{G-1}\frac{\{a\times b\}\times P(a,b)-\{u_\chi\times\mu_y\}}{\sigma_\chi\times\sigma_y} \tag{13.8}$$

13.2.2.6 Sum of Squares, Variance

It is said to be the calculation of dissemination of gray-level variations at certain distance d. It provides the essential weights on elements that are modified from higher value of $P(a,b)$, as provided in equation (13.9):

$$Variance = \sum_{a=0}^{G-1}\sum_{b=0}^{G-1}(a-\mu)^2 P(a,b) \tag{13.9}$$

13.2.2.7 Difference Entropy (DE)

$$DEnt = -\sum_{a=0}^{G-1}P_{x+y}(a)\log\left(P_{x+y}(a)\right) \tag{13.10}$$

where $P_\chi(a)$ is the ath entry of marginal-probability matrix attained by the inclusion of rows of $P(a,b)$, and $P_y(a)$ has been gathered by consolidating columns of $P(a,b)$. It is meant to be the estimation of histogram content as well as logical scores from two images. When the images are the same, then DE may be greater; else, it is lower.

13.2.2.8 Inertia

It showcases the decentralization of gray scales of an image, as given in equation (13.11):

$$Intertia = \sum_{a=0}^{G-1} \sum_{b=0}^{G-1} \{a-b\}^2 \times P(a,b) \tag{13.11}$$

13.2.2.9 Cluster Shade

The image becomes non-identical when maximum scale as well as cluster shade is presented as in equation (13.12):

$$Shade = \sum_{a=0}^{G-1} \sum_{b=0}^{G-1} \left\{a+b-\mu_\chi-\mu_y\right\}^3 \times P(a,b) \tag{13.12}$$

13.2.2.10 Cluster Prominence

This metric is represented as

$$Prom = \sum_{a=0}^{G-1} \sum_{b=0}^{G-1} \left\{a+b-\mu_\chi-\mu_y\right\}^4 \times P(a,b) \tag{13.13}$$

When images are non-identical, then prominences become higher.

13.2.2.11 Energy

The energy of a texture shows the texture consistency. Here, image is static when energy is 1, as shown in equation (13.14):

$$Energy = \sum_{a=0}^{G-1} \sum_{b=0}^{G-1} P(a,b)^2 \tag{13.14}$$

13.2.2.12 Homogeneity

This metric provides the value calculated from closeness of distribution elements in GLCM to GLCM diagonal. In case of diagonal GLCM, homogeneity becomes 1. A homogeneous image intends to generate co-occurrence matrix with the objective of reaching maximum and minimum $P[a,b]$'s. Next, $P[a,b]$'s are produced as given in equation (13.15):

$$Homogeneity = \sum_{a=0}^{G-1} \sum_{b=0}^{G-1} \frac{P(a,b)}{1+|a-b|} \tag{13.15}$$

13.2.2.13 Dissimilarity

It is defined as the measure of closeness between two communities, as implied in equation (13.16):

$$Dissimilarity = \sum_{a=0}^{G-1}\sum_{b=0}^{G-1} |a-b| P(a,b) \tag{13.16}$$

13.2.2.14 Difference in Variance

The inclusion of variations from intensity of intermediate pixel and adjacent neighborhood takes place, as given in equation (13.17):

$$Variance = \sum_{a=0}^{G-1}\sum_{b=0}^{G-1} (a-\mu)^2 P(a,b) \tag{13.17}$$

13.2.3 DNN-Based Classification

After GLCM-based feature extraction, the classification is performed on the feature values or vectors. Usually, classification is determined as an edge among the classes to label the classes depending on their evaluated features. Under the condition, DNN classifier is executed to classify the skin lesion images. A DNN is normally a feed forward networks and it can be an unsupervised pretraining method through greedy layer-wise training. At this point, the data flow from the input layer to the resultant layer with no looping function. A main benefit of DNN classification is that the probability of lost value is very small. The DNN method applies only one layer in unsupervised pretraining phase. A DNN assigns a classifier score $f(x)$ in the forecast time. All input data samples $x = [x_1, .., x_N]$ are forward pass. Typically, f is the function that contains an order of layers to calculate, which is expressed in equation (13.18):

$$Z_{ij=}x_i w_{ij}; Z_j = \sum_{\text{i}} Z_{ij} + b_j; X_j = g\left(Z_j\right) \tag{13.18}$$

where input of the layer is signified as x_i, the resultant layer is referred to as x_j, and w_{ii} are the method parameters, and $g(Z_j)$ appreciates the mapping or pooling function. A layer-wise significance propagation decays the classification result $f(x)$ with respect to relevance's r_i attributing to all input components x_i that gives to the classifier decision, as explained in equation (13.19):

$$f(x) = \sum_{\text{i}} r_i \tag{13.19}$$

where $r_i > 0$ denotes the positive data behind the classifier decision and $r_i < 0$ is the negative data of the classifier; or else, it is known as neutral data, while the relevance attribute r_i is computed utilizing equation (13.20):

$$r_i = \sum_j \frac{zij}{\Sigma_i zij}. \tag{13.20}$$

A DNN is capable to examine the indefinite feature coherences of input. A DNN gives hierarchical feature learning manner. Thus, the higher level features are taken from the minimum level features through a greedy layer-wise unsupervised pretraining information. So, a main purpose of DNN is to manage the difficult function that is illustrating higher level abstraction.

To optimize the parameters of the DNN, the GWO method is executed to fine-tune the parameters of the DNN model. A GWO is currently established swarm intelligence (SI) depending on the hunting system of gray wolf relations. The GWO method is provided in the several stages of GWO, as explained in the following sections.

13.2.3.1 Initialization of Grey Wolf Positions

It is function of a series of scale values $[s_1, s_2]$ as regarded in the range of [0, 1] in the initial and second modalities correspondingly. A count of iterations and population values are testing with several values utilizing trial and error model [18].

13.2.3.2 Social Hierarchy of Grey Wolf Family

Naturally, gray wolves are optimal in social management and are individual in a pack. A gray wolf pack has an average of 5–12 members. Hierarchical arrays of pack are separated into four levels: alpha (α), beta (β), delta (δ), and omega (ω). Generally, an initial level hierarchy α creates decisions considering hunting, sleeping place, time to wake, etc. A second-level position in pack is β, the optimal subordinates for α in pack stimulates. A least level arrange in gray wolf is ω, which plays the function of scapegoat. Besides, the ω wolves always contain submitting to every other hierarchical wolves. In a gray wolf relations, when the wolf is not an α, β, or ω, it is known as δ wolf. It contains submitting to α and β, however it controls the ω.

13.2.3.3 Encircling Prey

A gray wolf hunts the victim behind the assault and it surrounds a victim. The mathematical forms are provided in equations (13.21)–(13.24):

$$\vec{D} = \left|\vec{C} \cdot \overrightarrow{U_p}(t) - \vec{U}(t)\right| \tag{13.21}$$

$$\vec{U}(t+1) = \overrightarrow{U_p}(t) - \vec{A} \cdot \vec{D} \tag{13.22}$$

$$\vec{A} = 2\vec{a} \cdot \vec{r}_1 - \vec{a} \tag{13.23}$$

$$C = 2 \cdot \vec{r} \tag{13.24}$$

where t refers to the present iteration, $\overrightarrow{U_p}$ is the location vector of the victim, $\vec{U}$ signifies the location vector of gray wolf, $\vec{A}$ and $\vec{C}$ are coefficient vectors, and $\vec{D}$ is the direction vector. Here $\vec{r}_1, \vec{r}_2$ are arbitrary vectors in the series of [0, 1] and a linearly reduces from 2 to 0.

13.2.3.4 Hunting

Gray wolves contain the intellect for finding the place of its victim and neighboring. Generally, the hunt is directed by the guidance of α. The β and δ may be use in a pack. The α has the optimal information on victim location chased with β and δ. During GWO, the present iteration onward 3 optimally explore agents 5-007, β, and Δ to the next step, using each other agent in gray wolf relationships as gamma.

A mathematical form is provided as equations (13.25)–(13.27):

$$\vec{D}_\alpha = \left|\vec{C} \cdot \vec{U}_\alpha - \vec{U}\right|, \vec{D}_\beta = \left|\vec{C} \cdot \vec{U}_\beta - \vec{U}\right|, \vec{D}_\delta = \left|\vec{C} \cdot \vec{U}_\delta - \vec{U}\right| \tag{13.25}$$

$$\vec{U}_1 = \vec{U}_\alpha - \vec{A}_1 \cdot (\vec{D}_\alpha), \vec{U} = \vec{U}_\beta - \vec{A}_2 \cdot (\vec{D}), \vec{U}_3 = \vec{U}_\delta - \vec{A}_3 \cdot (\vec{D}_\delta) \tag{13.26}$$

$$\vec{U}(t+1) = \frac{\vec{U}_1 + \vec{U}_2 + \vec{U}_3}{3} \tag{13.27}$$

13.2.3.5 Attacking Prey

If victim prevents the going gray wolves, it terminates pursue by assaulting the victim. In this manner, the victim gray wolves require to reduce the value of a, where a is declined from 2 to 0 more than the course of iterations.

13.2.3.6 Search for Prey

A gray wolf generally explores depending on the position of the α, β, and δ. It diverges from all others for exploring the victims and converges for attacking the victim. For mathematical modelling of deviation, the arbitrary values superior than 1 or less than −1 are applied to make the explore agent for diverging from the victim. The C vector is regarded as the produce of obstacles for modelling victim naturally, which show in the chasing ways of wolves, and exactly reduce them from swiftly and calmly modelling victim. All explore agents update its distance from the victim.

13.3 EXPERIMENTAL VALIDATION

The experimental analysis of the GWO-DNN model is tested using ISIC dataset [19]. The dataset comprises a collection of 21 images under angioma, 46 images under nevus, 41 images under lentigo NOS, 68 images under solar lentigo, 51 images under melanoma, 54 images under seborrhoeic keratosis, and 37 images under BCC, respectively. Figures 13.2 and 13.3 show the sample set of images from the test dataset.

Figure 13.3 displays the qualitative results offered by the proposed model on the segmentation and classification of the skin lesion images.

Figures 13.4–13.6 illustrate the analysis of the results offered by the GWO-DNN model in terms of different performance measures. Figure 13.4 shows the analysis of the results offered by the GWO-DNN model in terms of sensitivity. The figure demonstrated that the SVM model appears as an ineffective performance, which has attained a least sensitivity of 82.78%. The MLP model has also exhibited inferior classifier results by attaining slightly

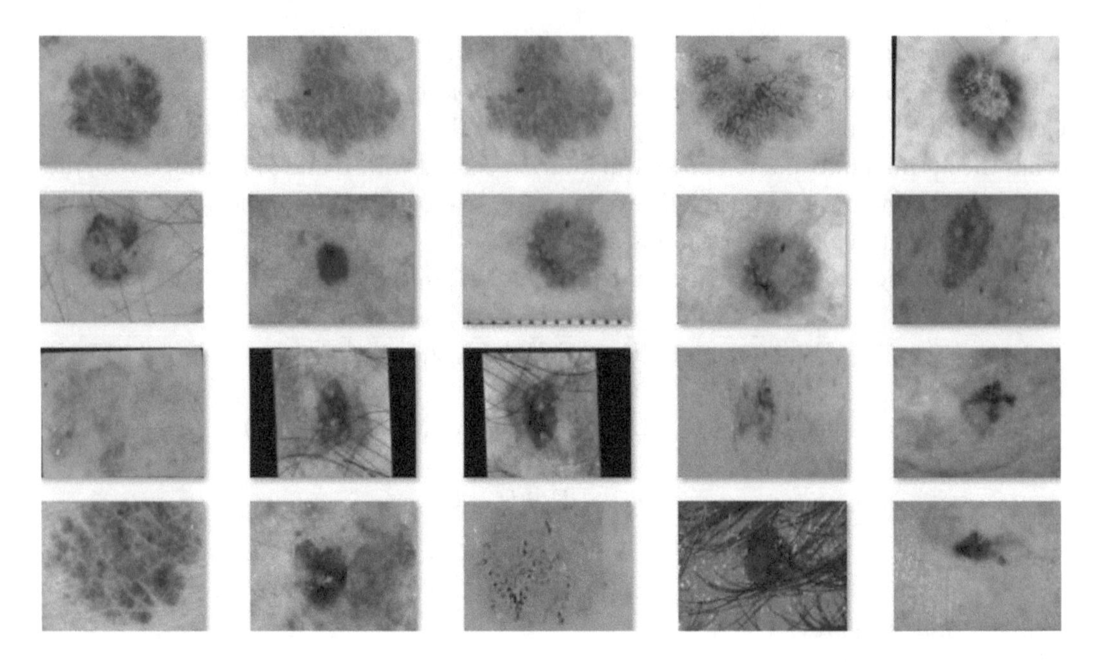

FIGURE 13.2 Sample images.

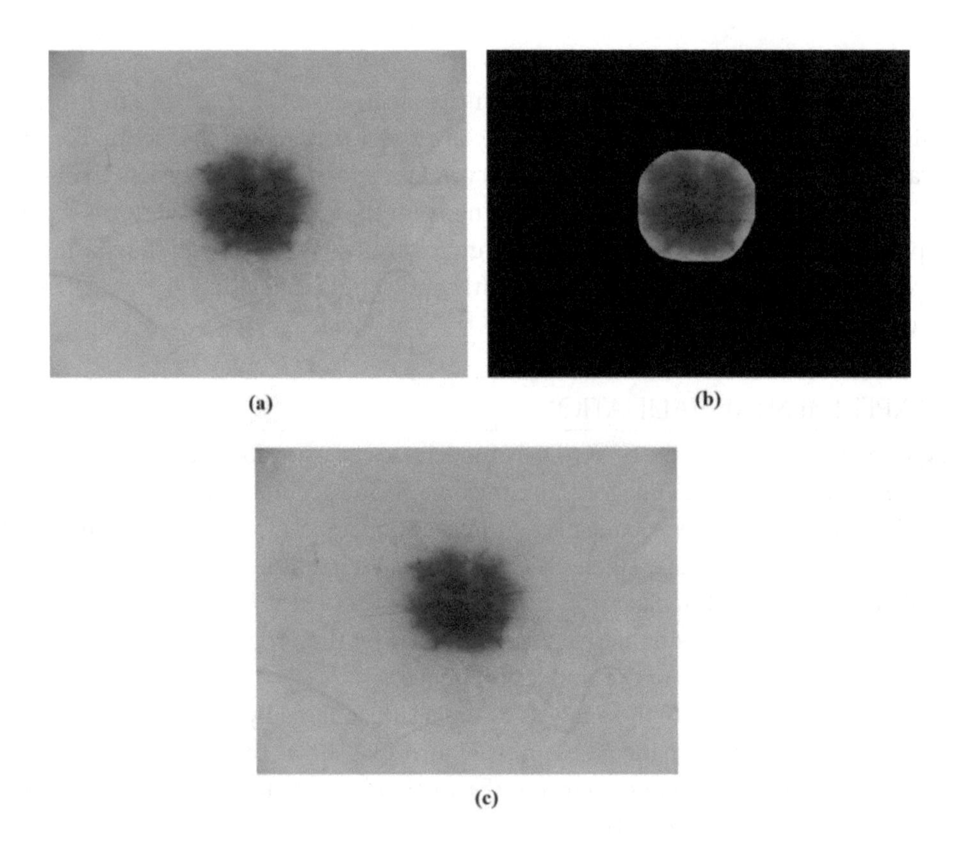

FIGURE 13.3 (a) Original image. (b) Segmented image. (c) Classified image.

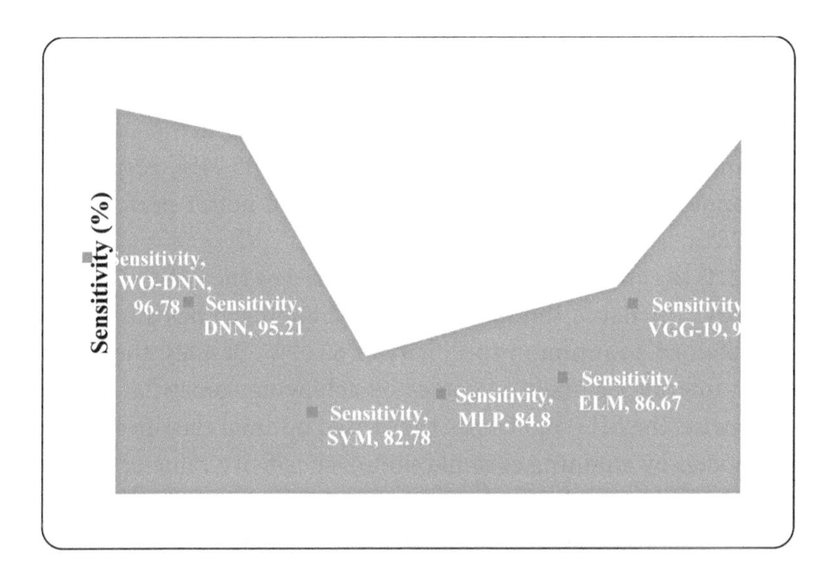

FIGURE 13.4 Sensitivity analysis of GWO-DNN model with existing methods.

FIGURE 13.5 Specificity analysis of GWO-DNN model with existing methods.

higher sensitivity value of 84.80%. Similarly, the ELM model has showcased better classification performance over the earlier models by achieving even higher sensitivity value of 86.67%. The VGG-19 and DNN models have also depicted the competitive results by obtaining high sensitivity values of 95% and 95.21%, respectively. Finally, the GWO-DNN model has performed well and attained better performance with the sensitivity of 96.78%.

Figure 13.5 depicts the analysis of the outcomes offered by the GWO-DNN method with respect to specificity. The figure shows that the SVM approach has an ineffective performance that has obtained a minimum specificity of 65.09%. Besides, the VGG-19 method has demonstrated inferior classifier outcomes by achieving somewhat higher specificity value of 68%. Likewise, the MLP model has showcased optimal classification performance over the earlier models by attaining even maximum specificity value of 70.55%. The ELM and DNN models have showed the competitive outcomes by attaining high specificity values of 74.80% and 75.90%, respectively. Finally, the GWO-DNN method has performed well and achieved optimal performance with the specificity of 76.54%.

Figure 13.6 illustrates the analysis of the results offered by the GWO-DNN method in terms of accuracy. The figure shows that the SVM model displayed an ineffective performance with a minimum accuracy of 70.43%. Furthermore, the MLP approach has performed inferior classifier results by obtaining slightly higher accuracy value of 74.32%. In the same way, the ELM method has exhibited better classification performance over the earlier models by obtaining even higher accuracy value of 77.99%. The VGG-19 and DNN methods have illustrated the competitive outcomes by achieving higher accuracy values of 81.20% and 81.65%, respectively. Finally, the GWO-DNN approach has exhibited well by obtaining optimal performance with the accuracy of 87.98%.

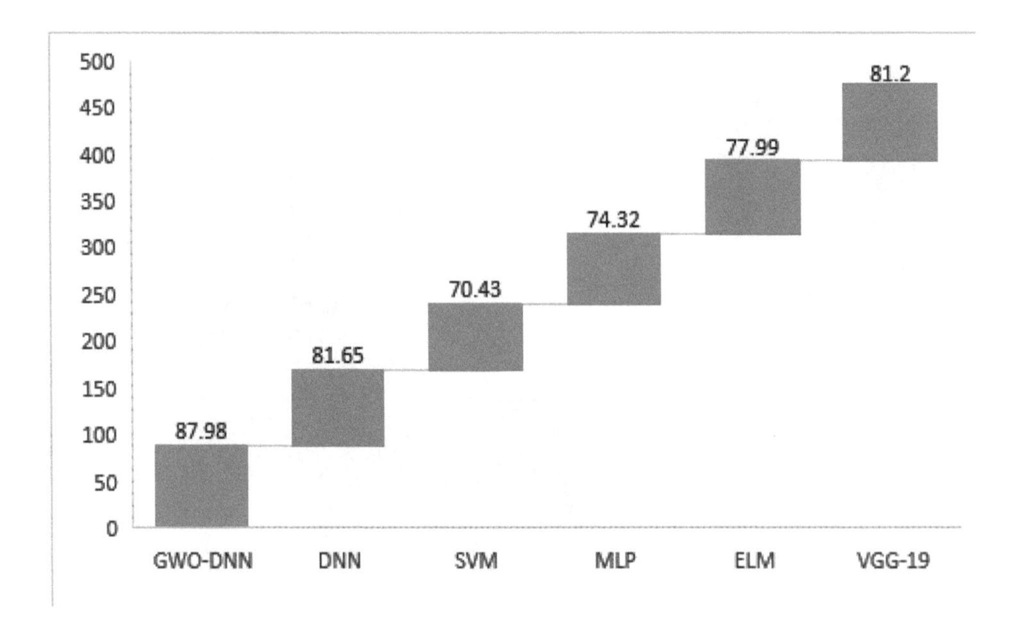

FIGURE 13.6 Accuracy analysis of GWO-DNN model with existing methods.

After examining the above-mentioned experimental analysis, it is evident that the GWO-DNN model is an excellent model over the compared methods. Therefore, it can be utilized in the real-time diagnosis of skin cancer using IoT devices.

13.4 CONCLUSION

This study has developed an effective automated skin lesion detection and classification using GWO-DNN model. The input images undergo several processes such as image pre-processing, segmentation, feature extraction, and classification. The GWO-DNN model performs the FCM-based image segmentation to identify the diseased portions in the dermoscopic image. Then, the GLCM technique is applied to extract the useful set of features. Finally, DNN optimized by GWO algorithm is applied as a classifier to categorize the different classes of skin lesion dataset. The experimental analysis of the GWO-DNN model is tested using ISIC dataset. The obtained experimental values confirmed the effective performance of the GWO-DNN model with the maximum sensitivity of 96.78%, specificity of 76.54%, and accuracy of 87.98%. The proposed model can be elaborated to the diagnosis of other diseases like COVID-19, diabetic retinopathy, etc.

REFERENCES

1. Stewart, B. W., & Wild, C. P. (2017). *World Cancer Report 2014*, World Health Organization: Geneva, Switzerland.
2. Gupta, S. & Tsao, H. (2017). Epidemiology of melanoma. In *Pathology and Epidemiology of Cancer*, Springer: Berlin/Heidelberg, Germany, pp. 591–611.
3. Rigel, D. S., Friedman, R. J., & Kopf, A. W. (1996). The incidence of malignant melanoma in the United States: issues as we approach the 21st century. *Journal of the American Academy of Dermatology*, 34(5), 839–847.
4. Lin, J., Han, S., Cui, L., Song, Z., Gao, M., Yang, G., ..., Liu, X. (2014). Evaluation of dermoscopic algorithm for seborrhoeic keratosis: a prospective study in 412 patients. *Journal of the European Academy of Dermatology and Venereology*, 28(7), 957–962.
5. Abbas, Q., Garcia, I., & Rashid, M. (2010). Automatic skin tumour border detection for digital dermoscopy using a new digital image analysis scheme. *British Journal of Biomedical Science*, 67(4), 177–183.
6. Barata, C., Ruela, M., Francisco, M., Mendonça, T., & Marques, J. S. (2014). Two systems for the detection of melanomas in dermoscopy images using texture and color features. *IEEE Systems Journal*, 8(3), 965–979.
7. Wollenberg, A., Günther, S., Moderer, M., Wetzel, S., Wagner, M., Towarowski, A., ..., Hartmann, G. (2002). Plasmacytoid dendritic cells: a new cutaneous dendritic cell subset with distinct role in inflammatory skin diseases. *Journal of Investigative Dermatology*, 119, 1096–1102.
8. Schmidt, E., Zillikens, D. (2010). Modern diagnosis of autoimmune blistering skin diseases. *Autoimmunity Reviews*, 10, 84–89.
9. Wang, J., Suárez-Fariñas, M., Estrada, Y., Parker, M.L., Greenlees, L., Stephens, G., ..., Howell, M.D. (2017). Identification of unique proteomic signatures in allergic and non-allergic skin disease. *Clinical & Experimental Allergy*, 47, 1456–1467.
10. Esteva, A., Kuprel, B., Novoa, R.A., Ko, J., Swetter, S.M., Blau, H.M., & Thrun, S. (2017). Dermatologist-level classification of skin cancer with deep neural networks. *Nature*, 542, 115.
11. Codella, N.C., Nguyen, Q.B., Pankanti, S., Gutman, D., Helba, B., Halpern, A., & Smith, J.R. (2017). Deep learning ensembles for melanoma recognition in dermoscopy images. *IBM J. Res. Dev.*, 61, 5(4–5), 15.

12. Schwarz, M., Soliman, D., Omar, M., Buehler, A., Ovsepian, S.V., Aguirre, J., & Ntziachristos, V. (2017). Optoacoustic dermoscopy of the human skin: tuning excitation energy for optimal detection bandwidth with fast and deep imaging *in vivo. IEEE Transactions on Medical Imaging*, 36, 1287–1296.
13. Li, Y. &Shen, L. (2018). Skin lesion analysis towards melanoma detection using deep learning network. *Sensors*, 18, 556.
14. Pathan, S., Prabhu, K.G., & Siddalingaswamy, P. (2018). Techniques and algorithms for computer aided diagnosis of pigmented skin lesions: a review. *Biomedical Signal Processing and Control*, 39, 237–262.
15. Chao, E., Meenan, C.K., & Ferris, L.K. (2017). Smartphone-based applications for skin monitoring and melanoma detection. *Dermatologic Clinics*, 35, 551–557.
16. Carmona, E.N., Sberveglieri, V., Ponzoni, A., Galstyan, V., Zappa, D., Pulvirenti, A., & Comini, E. (2017). Detection of food and skin pathogen microbiota by means of an electronic nose based on metal oxide chemiresistors. *Sensors Actuators B: Chemical*, 238, 1224–1230.
17. Behmann, J., Acebron, K., Emin, D., Bennertz, S., Matsubara, S., Thomas, S., Bohnenkamp, D., Kuska, M.T., Jussila, J., Salo, H., et al. (2018). Specim IQ: evaluation of a new, miniaturized handheld hyperspectral camera and its application for plant phenotyping and disease detection. *Sensors*, 18, 441.
18. Daniel, E., Anitha, J., Kamaleshwaran, K.K., & Rani, I., (2017). Optimum spectrum mask based medical image fusion using gray wolf optimization. *Biomedical Signal Processing and Control*, 34, 36–43.
19. https://www.isic-archive.com/

Index

Note: Page numbers in *italics* indicate figures and **bold** indicates tables in the text.

G

T

V

W

Y